STRATEGIC RETHINK

CHOICES FOR AMERICA IN A TURBULENT WORLD

James Dobbins | Richard H. Solomon | Michael S. Chase | Ryan Henry
F. Stephen Larrabee | Robert J. Lempert | Andrew M. Liepman
Jeffrey Martini | David Ochmanek | Howard J. Shatz

RAND CORPORATION

For more information on this publication, visit www.rand.org/t/rr1114

Library of Congress Cataloging-in-Publication Data is available for this publication.

ISBN 978-0-8330-9109-3

Published by the RAND Corporation, Santa Monica, Calif.

Original cover image from Fotolia; cover design by Peter Soriano

Preface

This book is the first of a series in which the RAND Corporation will explore the elements of a national strategy for the conduct of U.S. foreign policy in a fast-changing world. Here, we lay out the major choices the next administration will face both globally and in three critical regions. The initial chapters lay out alternatives for managing the world economy and the national defense, countering international terrorism, handling conflict in the cyber domain, and dealing with climate change. Subsequent chapters examine in more detail the choices to be faced in Europe, the Middle East and South Asia, and East Asia. The final section proposes broad strategic guidelines that can inform and guide these choices.

Later volumes will develop further particular aspects of such a national strategy, including national defense, alliances and partnerships, institutional reform of the American system for managing national security, climate change, surprise and the role of intelligence in reducing it, and the global economy.

This book should be of interest to defense and foreign policy decisionmakers, practitioners in the executive and legislative branches, analysts, the media, the staff and advisers to the 2016 presidential candidates, nongovernmental organizations, and others concerned with the role of the United States and other nations in advancing global security.

This project results from the RAND Corporation's Investment in People and Ideas program. Support for this program is provided, in part, by philanthropic contributions from donors and by the independent research and development provisions of RAND's contracts for the operation of its U.S. Department of Defense federally funded research and development centers. Special appreciation goes to the Hauser Foundation for its generous gift in support of the project and to Rita Hauser for encouraging RAND to undertake it.

Foreword

In 2013, amid war fatigue at home, tumult in the Middle East, and trepidations about Russian and Chinese intentions, I asked Ambassador Richard Solomon to lead a team of RAND colleagues in taking a fresh look at America's role in the world. With the United States heading into a presidential election campaign, I saw both a need to offer evidence-based, feasible policy options to inform the electoral debate and an opportunity to discuss larger strategic questions on which there is little consensus today.

What are America's international ambitions? What level of international engagement is the public prepared to support, and what can be sustained? How might the next president exercise international leadership in a tumultuous world, and to what end? And is there a coherent "grand strategy" for diplomacy and defense that would align the wide array of U.S. interests with the means to achieve them?

The result is our "Strategic Rethink" project. We pulled together some of our best minds to produce a guide for policymakers and citizens, educators and the media, on the most critical global choices and challenges that this president and the next will likely face—whether the public has yet focused on them or not.

This first volume was anchored by Ambassador James Dobbins, RAND's Distinguished Chair in Diplomacy and Security. It analyzes the choices, trade-offs, risks, and opportunities associated with various strategies for U.S. diplomacy, defense, and national security. Titled *Choices for America in a Turbulent World*, this book synthesizes the work of this team of experts on three critical regions—the Middle East and South Asia, Europe, and East Asia—as well as on national defense, international trade and economics, cybersecurity, and climate change. Its conclusions may surprise some. As you will see, Ambassador Dobbins does not believe the world is falling apart, even if the Middle East is. He sees opportunities to reinforce and expand the international order. The authors do not offer a single RAND game plan for national strategy, but rather examine the decisions that will be landing in the presidential inbox.

Subsequent volumes in the Strategic Rethink series, to be published over the coming months, will elaborate on key issues, including defense, international economic strategy, anticipating strategic surprise, making better use of intelligence, reforming the system of national-security decisionmaking, and managing the complex relationships with U.S. allies and partners.

Today, at every level of government, budgets are being sliced, squeezed, molded, and stretched; "doing more with less" is a common refrain. Yet no American president can have it all. The national debate about sensible budget decisions cannot succeed without rigorous analysis and unbiased appraisal of trade-offs. Any honest conversation about U.S. international engagement must match strategic options with the resources necessary to achieve them.

Through this series of publications, RAND aims to offer an informed perspective on these issues. As with most of our research and analysis, we have taken a multidisciplinary team approach, one that yields a diversity of views on the best strategies for tackling the key challenges facing the United States in the years to come. Over the course of this series, all of those voices will be heard.

I am pleased to share the analytic insights and the wisdom distilled from some of our most creative and distinguished colleagues.

Michael Rich
President and Chief Executive Officer,
RAND Corporation

Contents

Figures and Tables

Figures

Tables

Executive Summary

Foreign policy and national security seem likely to play a significant role in the 2016 presidential election campaign, and candidates from both parties will probably try to distinguish their approaches to these issues from that of the current administration. Recent events—most notably Russian aggression in Ukraine; the rise of the Islamic State in Iraq and Syria (ISIS);[1] and the continued proliferation of other violent extremist groups throughout the Middle East, South Asia, and much of Africa—have generated concern that current American responses are inadequate, yet no consensus has emerged around any alternative. Cybersecurity, climate change, and the increasing power of China are also growing as national preoccupations. While some criticize the Obama administration for weak and indecisive leadership, significant voices on both sides of the political spectrum argue for even greater restraint, lower resource commitments, and reduced engagement in addressing at least some of these issues.

Defining national strategy was easier when the country faced a single overarching threat. During the Cold War, it was possible to relate almost any endeavor to the genuinely existential competition with the Soviet Union. With the collapse of the Warsaw Pact and the Soviet Union, the expansion of Western values and institutions into former

[1] The organization's name transliterates from Arabic as *al-Dawlah al-Islamiyah fi al-'Iraq wa al-Sham* (abbreviated as Da'ish or DAESH). In the West, it is commonly referred to as the Islamic State of Iraq and the Levant (ISIL), the Islamic State of Iraq and Syria, the Islamic State of Iraq and the Sham (both abbreviated as ISIS), or simply as the Islamic State (IS). Arguments abound as to which is the most accurate translation, but here we refer to the group as ISIS or the Islamic State.

Soviet spaces became the focus of American policy. In the aftermath of September 11, 2001, the global war on terror became the organizing principle for American engagement with partners and against adversaries in every corner of the world. These oversimplifications led the United States down some costly and unnecessary paths, but such easy-to-grasp rationales nevertheless succeeded in mobilizing domestic and international support for strong action and important commitments.

Today, the United States faces no existential threat; rather, it confronts an unusually wide and diverse array of challenges. Russia has reemerged as an aggressor state. China has become more repressive at home and more assertive abroad. Al Qaeda has spawned offshoots and imitators more powerful and more radical than itself. Climate change has advanced, and predictions of climate-related disasters have become more ominous, more imminent, and more credible. Cyberspace has emerged as a new battleground between the forces of order and disorder. Expansion of international travel makes the emergence of new communicable diseases, such as Ebola, more dangerous. The past few years have been a reminder that stability is not the natural state of the international environment, that peace is not self-perpetuating, and that whole regions can descend suddenly into anarchy.

The world is more dangerous today than it was even just a few years ago, and the mounting chaos in the Middle East has fed wider anxieties. Many feel that the pace of technological change is quickening, that the international order is disintegrating, that power is shifting from national governments to individuals and nonstate actors, and that America's capacity to lead is waning. On balance, these portents of disaster are overstated. Certainly computers and the Internet are driving rapid change, but the overall pace is not more rapid or revolutionary than that following the introduction of electricity, radio, telephones, internal combustion engines, airplanes, and the atomic bomb. The Middle East is in turmoil, but the prevalence of inter- and intrastate conflict continues to decline, as do the casualties and destruction such conflicts produce. States are being challenged by terrorists and insurgent groups in the Middle East, as they once were in Southeast Asia, Latin America, and the Balkans, but governments are more capable today in those once-turbulent regions, and they remain so in Europe, East Asia, and North America. The Chinese economy has grown faster than that of the United States, but the United States has for many years been growing faster than Europe, Russia, and the rest of East Asia. Russia is misbehaving, but nothing on the scale of the former Soviet Union.

In 2016, presidential candidates will need to outline their intended responses to many of these challenges and, in the process, provide their vision of America's evolving role in a still fast-changing world. This publication is intended to highlight the most pressing foreign policy and security choices the next administration will face, to illustrate realistic alternatives available to the United States, to suggest several broad strategic orientations that might inform such decisions, and to lay out the elements of a national strategy that might offer the best prospects for success.

Choices for America

The United States faces a number of choices resulting from heightened global interdependence and its attendant vulnerabilities. The continued expansion of international trade, finance, travel, and communications has widened personal horizons, increased opportunity, promoted economic growth, and extended longevity for billions of people around the world. However, as we become more connected to the outside world, we become more affected by what occurs abroad. Climate change, for instance, is the direct result of global economic growth. Terrorists and criminals can mix with the millions of tourists and businesspeople who cross international borders every day. The communications revolution makes it easier for violent extremists to spread their ideology, recruit followers, and orchestrate attacks from great distances. As physical infrastructure becomes more dependent on digital controls, the possibility for catastrophic interference grows.

As the largest and fastest growing of the world's advanced industrial economies, the United States remains in a strong position to lead. Among the choices before it in the economic domain are

- whether to follow the conclusion of free-trade agreements with Europe and East Asia, assuming these are reached, with a return

to global trade liberalization or to continue to extend regional arrangements, and whether to include or exclude China from the latter
- how to structure existing and new multilateral arrangements for trade and finance either to accommodate an expanded Chinese role or to deal with the competitive institutions China is likely to form.

The U.S. defense budget is headed toward its lowest level, as a proportion of gross domestic product, in more than 50 years. The next administration and Congress will need to decide

- whether to continue improving U.S. capabilities to deter or defeat Chinese threats against allies and partners in East Asia, while also bolstering North Atlantic Treaty Organization (NATO) defenses in Europe, prosecuting the campaign against al Qaeda and ISIS, bringing the war in Syria to a satisfactory conclusion, and modernizing U.S. nuclear forces; four alternate levels of defense spending keyed to these objectives are presented.

In the domain of counterterrorism, the next administration must decide

- whether to combine aggressive attacks on terrorist networks with greater efforts in community outreach, counter messaging, counter radicalization, and other measures to reduce support for and participation in violent extremism
- whether to supplement air strikes against ISIS with participation in ground combat in Iraq and, eventually, Syria
- how narrowly or broadly to focus counterterrorist operations: only against groups that threaten the homeland or also against those that target U.S. friends and allies.

Similar issues will need to be faced in the cyber domain, including

- how best to balance privacy with security

- whether to seek stronger international norms regarding online behavior or rely solely on U.S. superiority to deter state and non-state attackers.

Climate change is the ultimate manifestation of interdependence and, thus, the test of the international community's ability to muster collective action. The next administration and Congress will need to decide

- whether the United States should lead or follow others in reducing carbon emissions
- whether to work toward that goal within existing multilateral mechanisms or rely on bilateral, regional, and new forums to do so.

The Obama administration's decision to rebalance U.S. engagement abroad away from Europe, the Middle East, and South Asia to East Asia was based on the assumption that the former regions were sufficiently stable to allow such a redirection of resources and attention. This has not proved to be the case.

In Europe, U.S. leaders face several interrelated issues:

- whether to seek a united but neutral Ukraine or see the country divided, with the larger part loosely associated with the European Union and NATO and the smaller dominated by Russia
- how far to go in isolating and penalizing Russia, given the need for its cooperation in other domains
- whether to deploy additional U.S. forces into or near the Baltic States to deter or defeat any Russian moves against these NATO allies.

In the Middle East and South Asia, the United States will face choices regarding ISIS, Iran, the Israeli-Palestinian divide, and Afghanistan:

- ISIS may be driven out of Iraq by the end of 2016, but as long as the civil war in Syria continues, that extremist group and others will continue to attract and inspire aspirant terrorists from all over the world. What combination of force and diplomacy should the United States employ to end that war?

- If a final nuclear accord with Iran is reached, how far should collaboration with Tehran go in Afghanistan and Iraq and in ending the civil war in Syria?
- Should the United States accept the status quo in Israeli-Palestinian relations, make more-substantial efforts to press both sides toward a two-state solution than have been tried in the past, or begin exploring more seriously the one-state alternative?
- Should all U.S. and NATO forces pull out of Afghanistan by the end of 2016, or should a small residual military presence remain?

In East Asia, the main choices relate to China and the Democratic People's Republic of Korea (North Korea):

- Should the United States accommodate China's growing power by affording it greater weight in the management of multilateral trade and financial institutions, or seek to limit that influence to retain Western dominance in these areas?
- More generally, what balance of containment and engagement should the United States employ in dealing with an increasingly assertive China and increasingly anxious U.S. partners in the region?
- Similarly, what combination of isolation and engagement should the United States employ in dealing with North Korea?

Assessing and Accepting Risk

Many of the above choices come down to the trade-off between expenditure and risk. The United States can spend more on defense or risk strategic setbacks in Europe, East Asia, and the Middle East. It can spend more on reducing carbon dioxide emissions or risk the many negative consequences of climate change. As long as the risk is not existential, recovery remains possible, albeit perhaps at much greater expense. The United States was unprepared for World War II, for the

Korean War, for Saddam Hussein's seizure of Kuwait in 1991, and for 9/11, but it recovered from each. The size and power of the United States relative to the rest of the world provide it with unique resiliency. In such cases, strategic decisions may come down to a choice between spending more now or possibly having to spend a lot more later on.

It thus makes sense not only to examine desirable objectives but also minimally acceptable outcomes. In Europe, the United States might like to see a unified Ukraine moving toward NATO and European Union membership but might be able to accept a divided Ukraine with the larger part linked to Europe but forswearing NATO membership. In the Middle East, the United States would like to see a peaceful Syria under a moderate pro-Western government but, at this point, could probably live with a peaceful Syria under almost any government actually able to impose control and maintain order. In East Asia, the United States faces little difficulty helping protect its treaty allies against actual invasion and occupation. The sources of friction with China tend, in some cases, to be issues in which the United States has no direct interest. Absent some major miscalculation, the risks are not existential for the United States or its allies. Some nontreaty partners or potential partners are more vulnerable. So, there are important cost and risk equations to be evaluated when seeking to contain China.

This is not to suggest a race to the lowest acceptable outcome but to note that the cost-benefit ratio associated with doing better than the acceptable minimum needs to be part of the decision process. Declaratory policy will naturally set out desired outcomes; actual strategy needs to leave room for bearable outcomes that fall short of declared goals. Strategic failure will ensue when resources committed prove insufficient to reach even minimally acceptable goals.

Values and Interests

Like its predecessors, the next administration will want to ground its national strategy in American values and interests. The ultimate test of leadership, after all, is the ability to inspire followers. Washington needs to enunciate policies that the public will support, partners will join, and adversaries will respect.

At a certain level of abstraction, this is easy enough to achieve. The United States values democracy and free markets and is interested in peace, international collaboration, and expanding trade. It is easier to collaborate with established democracies and to trade with free-market economies than it is to do so with authoritarian governments and closed economies. Therefore, our values and interests cohere.

In reality, promoting values can be rather more complicated. Nondemocratic regimes resist and resent efforts to remake them in our image and will sometimes withhold collaboration on otherwise shared interests as a result. Freedom, democracy, and human rights may be universally applicable, but, much as we would wish otherwise, they are not universally attractive, particularly within conservative societies, where gender inequality and authoritarian rule are sanctioned and even enforced by religious authority. Finally, as the aftermath of the Arab Spring has demonstrated, there are worse things than a cooperative authoritarian government, such as an uncooperative and even more authoritarian government, as in today's Egypt, or anarchy and bloodshed, as in today's Yemen, Libya, and Syria.

In practice, therefore, there is sometimes tension between promoting democracy and human rights and advancing security and economic interests abroad. This requires case-by-case assessment of the local receptivity to such efforts; the cost in terms of other issues of pressing too hard; and the likelihood, if change comes, that it will move in the right direction.

Keeping Pace with Change

As noted previously, there is a common perception that the world is changing with unprecedented speed, making it difficult for U.S. policymakers to keep up. Certainly, information flows more quickly and more widely than ever, allowing and even requiring rapid responses to distant events. Moore's law, the observation that computing power will double annually, is sometimes applied more broadly to suggest a comparable acceleration of geopolitical developments. Events in the Arab world are cited to illustrate this thesis.

However, in terms of shifts in global power balances, prior periods have been even more marked by rapid change. World War I brought the collapse of the Ottoman and Austrian empires and the creation of more than a dozen new countries. During the two decades after World War II, control over more than one-half the world's surface and population shifted radically as dozens of nonstate actors—then known as liberation movements—seized power and set up new regimes, both in established states, such as China, and throughout the colonial world. Even without the benefit of television and the Internet, contagion and imitation from one society to another were quite rapid throughout this period. Change occurred at an even greater speed during the first two years of the George H. W. Bush administration, with the unification of Germany, the disintegration of the Warsaw Pact, and the collapse of the Soviet Union. Again, there was broad contagion, with similar political changes occurring throughout Eastern Europe and with 15 new states emerging from the former Soviet Union. Further, with the end of the Cold War came the quick settlement of a number of civil wars in Africa, Asia, and Latin America that had been stimulated by superpower competition.

Most of these earlier geopolitical shifts benefited the United States—unlike current developments in the Arab world. Not surprisingly, people seem more likely to notice the pace of change for the worse, as opposed to the pace of change for the better. Nevertheless, it would be difficult to maintain that the distribution of power among states is changing more quickly today than it did after 1918, 1945, or 1989.

This leaves the issue of the distribution of power between states, on the one hand, and nonstate actors, including individuals, on the other. Are states losing their grip? Is power devolving downward? An oft-heard concern is that the interstate system established in the Treaty of Westphalia in 1648 is unraveling. But most governments in Europe, East Asia, and the western hemisphere have not experienced diminished capacity. Africa has long been home to a number of failed and failing states, but the problems there are no more acute today than at any time since the decolonization of that continent some 60 years ago.

What is new and disturbing is the fragility of Arab states. Several have descended into civil war. The rest are deathly afraid of doing so, leading their governments to take extreme and often ill-considered measures to counter what they regard as the forces of dissolution.

Region-wide upheavals are by no means a new phenomenon. In the 1950s, anticolonial revolts, usually employing terrorism, created dozens of new countries throughout Africa and Asia. In the 1960s and 1970s, all of Southeast Asia was engulfed in conflict. In the 1980s, Latin America and sub-Saharan Africa experienced multiple civil wars. At the end of the Cold War, the Balkans exploded. Throughout these periods, most authoritarian Middle Eastern regimes exhibited a durability based on rigid resistance to change, which has contributed to their current fragility.

Another concern is that even well-functioning states are losing ground as power is dispersed downward and outward. The continued expansion of international trade, finance, travel, and communications has exposed vulnerabilities even as it has widened horizons, increased opportunity, and lifted people around the world out of poverty. Terrorists and criminals can mix with the millions of tourists and business people who cross international borders every day, but security agencies also have new, more-powerful tools to track and impede their movements. Contagious disease can spread more rapidly, but resources to contain it can also be mobilized and dispatched more quickly. The communications revolution empowers individuals and states alike. Violent extremists can more easily spread their ideology, recruit followers, and orchestrate attacks, but security authorities can more easily collaborate to foil these attempts. As physical infrastructure becomes more dependent on digital controls, the possibility for catastrophic interference grows, requiring ever-higher levels of digital safeguards.

The Arab world aside, states are not fragmenting—nor are most of them losing ground to individuals or groups, malign or otherwise. Technology is neutral and can be used both to challenge and extend state authority. Admittedly, there is a race between the forces of order and disorder in all these domains, but it is not one that effective states are predestined to lose.

Overlearning the Lessons of the Past

It took the United States a generation to get over the lost war in Vietnam; it can ill afford to wait another generation before recovering from the setbacks in Iraq and Afghanistan and the disappointing results of the Arab Spring. This is not to say that policymakers do not need to reflect on and absorb the lessons of the past decade, but these lessons are not that nation-building never works, that counterinsurgency is always too expensive, or that democracy promotion is ineffectual and potentially counterproductive.

Tomes have been written about the mistakes made in Afghanistan and Iraq during the Bush administration's first term. The United States went into both countries determined not to engage in nation-building, so it should surprise no one that its record in this regard disappoints. In both cases, the administration grossly underestimated the resources needed to stabilize these societies, eventually deploying the needed assets only years later, by which time large, well-organized resistance movements had emerged.

The lesson to be drawn from Afghanistan and Iraq is not that nation-building does not work but rather that it can be very expensive and time consuming. There are more than a dozen countries around the world today that are at peace because U.S. troops—or NATO, European, United Nations, or African Union forces—intervened to end a civil war, provided security to the population, oversaw elections, helped install new governments, and stayed around long enough to make sure the new regime took hold. Few of these societies are prosperous, well governed, or fully democratic, but they are more prosperous, better governed, and more democratic than before. Most importantly, they are at peace with themselves and their neighbors, which was the prime objective of the interventions in the first place.

A second lesson already evident in Iraq and in Libya is that forced regime change that is not followed by successful stabilization operations can actually create a situation worse than the one the original intervention was intended to correct.

The objective of a stability operation is to prevent the renewal of conflict. Insurgency is what one gets if the stability operation fails—

as happened almost immediately in Iraq and after several years in Afghanistan—or is never attempted, as in Libya. At that point, one must either counter the insurgency or allow it to prevail. Obviously, it is better to leave this to indigenous forces—assuming they exist—but the United States will not be able to help other regimes carry out successful counterinsurgency operations if it lacks the expertise and capacity to do so itself.

Democratization is not a binary condition. Some societies moved rapidly from authoritarian to democratic rule. In others, that process was much more gradual. Since the end of World War II, and particularly since the end of the Cold War, dozens of countries have moved away from the authoritarian camp, and now democracy is the dominant form of government throughout the world. There has been some limited regression over the past decade, with several Arab societies having transitioned from authoritarian governments to none at all in the wake of the Arab Spring. Rather than giving up on the process, however, Americans need to temper their expectations and work to promote the foundations of democratic government—civil society, rule of law, growth of the middle class—so that, when future upheavals occur, as they inevitably will, the results will be more positive. We also need to recognize that almost any government is better than none.

U.S. National Strategy

Since becoming the world's most powerful nation in the middle of the last century, the United States has labored to build a rules-based global system dominated by market democracies and dedicated to promoting free trade and the peaceful settlement of disputes. This new order emerged from the Truman Doctrine, the Marshall Plan, and the reconstruction of Germany and Japan. It was bolstered by the creation of the United Nations, NATO, the World Bank, the International Monetary Fund, and the World Trade Organization. It eventually produced the end of colonialism, the reunification of Germany, the liberation of Eastern Europe, the peaceful dissolution of the Soviet empire, the extension of democracy to 144 countries,[2] and the greatest advances in global longevity and poverty reduction in human experience. Preserving and extending this world order should remain the core objective of U.S. policy—first, because it provides the best environment for the United States' own security and prosperity and, second, because if the United States does not lead, no one else will. Defining a national strategy toward this end requires the identification of ends, ways, and means.

Ends

American interests and values will be best served in a world in which states adhere to established norms of behavior that ensure peace and promote prosperity. Importantly, these same states must remain capable and willing to ensure that their citizens observe such norms. Threats to this order are thus posed by both states that flout the rules and states that prove incapable of enforcing them. Any unwillingness or incapacity to exercise the responsibilities of sovereignty becomes all the more dangerous in a world ever more closely knit by trade, travel, and communications, a world in which mere physical proximity is no longer the prime factor in determining the source of serious threats.

Today's state-based threats to international order come primarily from Russia, Iran, North Korea, and (potentially) China. Russia's aggression against Ukraine needs to be made costly enough to discourage repetition or imitation. China needs to be deterred from any similar behavior. Iran must be prevented from acquiring nuclear weapons. North Korea must be prevented from exporting its nuclear capacity and dissuaded from attacking its neighbors. China's growing power needs to be channeled constructively. One can pose and even pursue more-ambitious goals, such as turning back Russian aggression, dismantling Iran's nuclear infrastructure, denuclearizing Korea, or preventing an expansion of Chinese influence, but the costs and risk associated with these more-ambitious aims may be prohibitive.

---

[2] Freedom House finds 88 countries free and 59 partly free, representing a decline over the past decade but still much higher than in any other era. See Freedom House, *Freedom in the World 2014*, Washington, D.C., 2014.

Today's nonstate threats come principally from violent extremist groups based in the Middle East, North Africa, and South Asia. Eliminating these groups entirely is beyond U.S. capacity, but they should at least be suppressed to the point that they no longer hold territory and govern large populations, are no longer able to do great damage at great distance, and no longer recruit and inspire large numbers of individuals from within Western and other societies. Ending the civil war in Syria, even at the expense of dealing with Assad, is probably the single most important step toward suppressing the most virulent of these groups and diminishing their ability to attract new recruits and inspire imitators. This will require engaging Iran while also maintaining existing U.S. partnerships and alliances.

The defensive aspects of American strategy must be to punish and deter state-on-state aggression while suppressing violent nonstate extremist movements that threaten U.S. citizens and those of our friends and allies. For these efforts to have any lasting effect, the United States also needs to sustain and extend a rules-based international order founded on states that are willing to observe certain norms of behavior and able to ensure that their citizens do likewise. A major goal of U.S. policy should therefore be to improve the capacity of international institutions to channel collective action; to strengthen the capacity of individual states to engage effectively in such action; and to further develop international norms in new areas of vulnerability, most notably in the cyber and climate domains, where an adequate body of generally accepted rules are currently lacking.

Ways

The ways of dealing with threats from strong states are conceptually well developed, if expensive and demanding in application. These include diplomacy, economic pressure, deterrence, containment, and collective defense.

The ways of dealing with threats emanating from weak states include counterterrorism and counterinsurgency operations combined with state- and nation-building efforts.

Counterterrorism and counterinsurgency campaigns can disrupt extremist movements, but only the establishment of states willing to adhere to and capable of enforcing international norms will allow such campaigns to be concluded. State- and nation-building operations are long-term, resource-intensive enterprises, but at least they have a defined end state. Counterterror operations do not.

In today's hyperlinked world, few challenges are amenable to unilateral responses. American involvement will be essential to the resolution of the many challenges outlined in this book, but it is seldom sufficient. Slowing climate change is only the most extreme example. Deterring or turning back aggression almost always requires collective defense. State- and nation-building efforts require support from neighbors and near neighbors, from the societies that have the most at stake and possess the most influence by reason of their proximity and their commercial, criminal, religious, and cultural connections. Partnerships are essential. Coalitions are the norm. One cannot afford to be too choosy about the company one keeps. Russia is needed to help prevent Iran from acquiring nuclear weapons. China is needed to restrain North Korea. Iran is needed to fight ISIS. Both Russia and Iran will be needed to end the war in Syria. Stemming climate change will require almost global efforts. Partnerships in this era are not just about friends confronting enemies.

More government may not be the answer to America's domestic problems, but more and better government abroad is certainly essential to meet most of its external challenges. State-building will remain an unavoidable mission for the United States in Europe, the Middle East, and South Asia, as well as in Africa. This need not always involve military interventions, nor must the United States be the principal source of military manpower and economic assistance. Where regional states have been willing to take the lead, as in Africa, U.S. advice and financial assistance may be all that is necessary. Building a functional Ukrainian state and continuing to stabilize the Balkans will be tasks primarily for Europe. But the United States should not base its defense planning and size its own military establishment on the premise that it is permanently out of the nation-building business. There is nothing in our history or our prospective future that supports such an assumption.

Means

As a proportion of national income, the U.S. defense budget is headed for its lowest level in some 50 years. Yet with the continued rise of China, the emergence of ISIS, and intensified Russian belligerence, the United States now faces the need to fight, or at least deploy forces sufficient to fight, on three fronts against at least three different opponents. The current budget and envisaged force structure are inadequate to this task, as successive U.S. secretaries of defense have acknowledged.

During the Cold War, the United States sized and organized its military establishment to be able to fight and win two major wars at once, one in Asia and one in Europe, against two different opponents. In recent decades, this requirement has been nominally sustained, but the scale of each envisaged conflict has been reduced, reflecting both a diminished threat and reduced capacity. The current standard is to defeat one regional adversary in a large-scale, multiphased campaign and deny the objective of, or impose unacceptable costs on, a different aggressor in another region.[3] Additionally, discouraged by the results of its efforts in Iraq and Afghanistan, the Obama administration has decided not to require U.S. armed forces to be ready to conduct large-scale, protracted stability or counterinsurgency operations and has cut the size of the Army and Marine Corps accordingly. Meanwhile, many modernization programs have been slowed or truncated, and readiness has eroded due to reduced levels of defense funding.

These trends are clearly incompatible with the three-theater challenge the United States now faces. The current international environment does not permit the United States to continue to transfer its time, attention, and national resources from West to East. Further military reductions from Europe, the Middle East, and South Asia to East Asia are not prudent. The current decline in the U.S. defense budget—and the national security budget more generally—must be slowed, and all three theaters need to be adequately resourced.

---

[3] Department of Defense, *Sustaining U.S. Global Leadership: Priorities for 21st Century Defense*, Washington, D.C., January 2012.

Conclusion

Within American society, the predominant but never-uncontested tendency for most of the period since World War II has favored American leadership in the sustainment and expansion of a rules-based international order promoting free trade and the peaceful settlement of disputes, combined with a willingness to commit diplomatic, economic, and military resources to develop and help enforce such norms.

A more-hawkish camp has been equally committed to American global leadership and even more ready to devote resources to that effort but places somewhat less emphasis on multilateralism and the further elaboration of enforceable norms of behavior.

A third group feels the United States is overextended abroad, does not believe it needs to assume the dominant responsibility for addressing every challenge to global order, and would like to reduce the overall level of its overseas commitments. On the political right, this group includes small-government conservatives and libertarians. On the left, its supporters include those who prefer "nation-building at home," generally oppose the use of force, and resist further globalization.

There remains, nevertheless, a sizable constituency in Congress, as in the public, in favor of a well-resourced foreign and security policy agenda, but it spans both parties and dominates neither. This constituency can prevail only if there is willingness on both sides to cross party lines and support sensible policies. As long as the two parties remain dug in behind the partisan ramparts of "no new taxes" and "no cuts to entitlements," the United States' external problems will mount as new challenges pile on top of old ones that have not been addressed. As a result, the global order will indeed begin to erode, and fundamental American interests will suffer.

The United States, as the world's most powerful nation, should continue to take the lead in sustaining and extending a rules-based international order. It should promote the development of new norms in domains where these do not yet exist, such as cyber and climate management. States are the essential building blocks in any such system. Challenges come from strong states that break the rules and weak ones that cannot enforce them. Both these challenges need to be addressed.

A focus on defense, deterrence, and dissuasion is essential, but it is not enough. State capacity needs to keep pace with the growing capacity for disruption of individuals and groups. The most successful eras of American statecraft have been periods of construction: the birth of new institutions, the reconstruction of shattered nations, and the establishment of new norms for international behavior. The United States needs to combine its defense of existing institutions and norms with a rededication to such a positive agenda and to commit itself to providing the necessary resources.

Acknowledgments

The authors would like to thank Richard Solomon and Andrew Hoehn for leading RAND's Strategic Rethink project, of which this is the initial product; RAND president and CEO Michael Rich, who initially conceived of the project; Brian Katulis, Daniel Chu, and Mike Mazar for their very helpful reviews; and Sonni Efron for her many editorial insights and suggestions. Also, special thanks are due to Rita Hauser and the Hauser Foundation, without whose encouragement and financial support this project would not have been possible.

CHAPTER ONE

Introduction

> The United States finds itself in a paradoxical situation. By any standard of national capacity, we are in a position to achieve our objectives and to shape international affairs.
>
> Yet as we look around the world, we encounter upheaval and conflict. The United States has not faced a more diverse and complex array of crises since the end of the Second World War.
>
> —*Henry Kissinger*[1]

Foreign policy and national security seem likely to play significant roles in the 2016 presidential election campaign. Candidates from both parties will probably try to distinguish their approaches to these issues from that of the current administration. Recent events—most notably Russian aggression in Ukraine; the rise of the Islamic State in Iraq and Syria (ISIS);[2] and the continued proliferation of other violent extremist groups throughout the Middle East, South Asia, and much of Africa—have generated concern that current American responses

---

[1] Henry A. Kissinger, *Global Challenges and U.S. National Security Strategy*, testimony before the U.S. Senate Armed Services Committee, Washington, D.C., January 20, 2015.

[2] The organization's name transliterates from Arabic as *al-Dawlah al-Islamiyah fi al-'Iraq wa al-Sham* (abbreviated as Da'ish or DAESH). In the West, it is commonly referred to as the Islamic State of Iraq and the Levant (ISIL), the Islamic State of Iraq and Syria, the Islamic State of Iraq and the Sham (both abbreviated as ISIS), or simply as the Islamic State (IS). Arguments abound as to which is the most accurate translation, but here we refer to the group as ISIS or the Islamic State.

are inadequate, yet no consensus has emerged around any alternative. Cybersecurity, climate change, and the increasing power of China are also growing national preoccupations. While some criticize the Obama administration for weak and indecisive leadership, there are also significant voices on both sides of the political spectrum arguing for even greater restraint, lower resource commitments, and reduced engagement in addressing at least some of these issues.

In 2016, presidential candidates will need to outline their intended responses to many of these challenges and, in the process, provide their vision of America's evolving role in a fast-changing world. This book is intended to highlight the most pressing foreign and security choices the next administration will face, to illustrate realistic alternatives available to the United States, to suggest several broader strategic orientations that might inform such decisions, and to lay out the elements of a national strategy that offers the best prospects for success.

Defining national strategy was easier when the country faced a single overarching threat. During the Cold War, it was possible to relate almost any endeavor to the genuinely existential competition with the Soviet Union. With the collapse of the Warsaw Pact and the Soviet Union, the expansion of Western values and institutions into former Soviet spaces became the focus of U.S. policy. In the aftermath of September 11, 2001, the global war on terror became the organizing principle for American engagement with partners and against adversaries in every corner of the world. These oversimplifications led the United States down some costly and unnecessary paths, but such easy-to-grasp rationales nevertheless succeeded in mobilizing domestic and international support for strong action and important commitments.

Today the United States faces no existential threat but, as former Secretary of State Henry Kissinger noted, the country does confront an unusually wide and diverse array of challenges. Russia has reemerged as an aggressor state. China has become more repressive at home and more powerful abroad. Al Qaeda has spawned offshoots and imitators more powerful and even more radical than itself. Global climate change has advanced, and predictions of climate-related disasters have become more ominous, more imminent, and more credible. Cyberspace has emerged as a new battleground between the forces of order and disorder. Expansion of international travel makes the emergence of new communicable diseases, such as Ebola, more dangerous. The past few years have been a reminder that stability is not the natural state of international environment, that peace is not self-perpetuating, and that whole regions can descend suddenly into anarchy.

Alternative Models

While actual national strategy may emerge as the sum of many separate and largely unrelated decisions, any administration will seek to define its priorities and lay out its policies in a coherent fashion to both guide its own efforts and secure domestic and foreign support. The academic literature offers several competing templates for assigning such priorities.

The realist perspective emphasizes the centrality of states (particularly great powers), the inevitable competition between them for influence and power, and the primacy of security and economic objectives as the focus of this competition. This school tends to envision national interests not as what a given population may actually be interested in but rather as goals toward which all nations tend to strive.

The neoconservative school shares with realists an emphasis on the importance of hard power but with an inclination to employ it toward softer—that is, more value-based—objectives, most notably democracy promotion.

An opposing school advocates for the United States to adopt the role of an "offshore balancer," in which it would employ its political and economic weight, along with its maritime military capacity, to maintain regional equilibriums without engaging U.S. assets too heavily. This is a "no boots on the ground" variant of realism.

The Wilsonian tradition, named after President Woodrow Wilson, shares with neoconservatism the evangelizing goal of democratization but sees this arising more from good example and willing emulation than from coercion. It also places greater emphasis on multilateralism, international law, and thus regional and global institutions than the other schools.

The national security strategies of recent administrations contain traces of all these schools. In the immediate aftermath of the Cold War, there was a general expectation that the rest of the world would naturally gravitate to the only societal model left standing, that of liberal Western democracy. The national strategy during President Bill Clinton's first term thus set as its core goal expanding this democratic community through the power of example and the deployment of positive incentives. Accordingly, the administration championed the expansion of both the European Union (EU) and the North Atlantic Treaty Organization (NATO). Later in the decade, this Wilsonian vision took on more forceful overtones as the administration sought to protect endangered populations and promote democracy in the Balkans, albeit still within a multilateral framework.

President George W. Bush campaigned on a realist platform (no nation-building) but moved in a neoconservative direction in reaction to the 9/11 attacks.

President Barack Obama has pursued a policy of retrenchment abroad and nation-building at home, a tentative form of offshore balancing.

The realist model adapts itself fairly well to East Asia, a region of strong states governed by regimes that tend toward realist, value-free external policies and without significant insurgencies or terrorist movements.

International politics in Europe today are much more value-laden. The continent's problems derive as much from weak states on its periphery as from competition among peers. Here, the choices largely lie between realism, which would tend to leave these weak peripheral states to their own devices and thus effectively allow them to be drawn into the Russian orbit, and Wilsonian and neoconservative approaches, which would seek to bring these outlying states into the Western economic and security perimeter.

For much of the post–World War II era, the Middle East presented a classic opportunity for American offshore balancing. For decades, Iraq and Iran contained each other effectively, with minimal need for U.S. involvement. As a result, the United States could maintain acceptable equilibriums between Arabs and Israelis and among

Arab states with periodic, usually minimal and short-term applications of U.S. military force. Saddam Hussein's 1991 invasion of Kuwait drew the Unites States more deeply into the region. The American invasion of Iraq in 2003 destroyed the Iraqi-Iranian equilibrium. At present, there does not seem any regional balance to maintain.

Beyond these theoretical constructs, there are three broad strategic tendencies that compete for support within the American body politic.

The predominant but never-uncontested view for most of the period since the Second World War has favored well-resourced internationalism—that is, American leadership in the sustainment and expansion of a rules-based international order promoting free trade and the peaceful settlement of disputes, combined with a willingness to commit diplomatic, economic, and military resources to develop and help enforce such norms.

A more-hawkish camp places somewhat less emphasis on multilateralism and tends to be even more ready to employ economic and military coercion in support of U.S. security interests.

The third group feels the United States is overextended abroad, does not believe it needs to assume the dominant responsibility for addressing every challenge to global order, and would like to reduce the overall level of its overseas commitments. On the political right, this group includes small-government conservatives and libertarians. On the left, its supporters include those who prefer "nation-building at home," generally oppose the use of force, and resist further globalization.

Disappointment with Afghanistan, Iraq, and the Arab Spring has left all three of these camps more skeptical about both nation-building and democracy promotion.

Assessing and Accepting Risk

This study works toward the development of a national strategy by first examining the actual policy choices that the next administration is likely to face. Many of these choices come down to the trade-off between expenditure and risk. The United States can spend more on defense or risk strategic setbacks in Europe, East Asia, and the Middle East. The United States can spend more on reducing carbon emissions or risk the many negative consequences of climate change. As long as the risk is not existential, recovery remains possible, albeit perhaps at much higher expense. The United States was unprepared for World War II, for the Korean War, for Saddam Hussein's seizure of Kuwait in 1991, and for 9/11, but it recovered from each. The size and power of the United States relative to the rest of the world provide it with unique resilience. In such cases, strategic decisions may come down to a choice between spending more now or possibly having to spend a lot more later on.

Many of the alternatives set out in this book are of this nature. As was the case with the loss of the Vietnam War, it is likely that the United States could recover from setbacks arising from aggressive Russian behavior in Europe, Chinese behavior in East Asia, or violent extremist groups from wherever. There would be costs to America's geopolitical position, its credibility, and its alliance relationships. Recovery would be expensive and would bring with it new risks. It makes sense to hedge against these costs and risks by making current investments, but, given finite resources, it also makes sense to accept some level of risk. In retrospect, for instance, it may seem that President Obama miscalculated in withdrawing all U.S. troops from Iraq in 2011, but this may have been a reasonable cost-benefit calculation at the time.

It thus makes sense to examine not only desirable objectives but also minimally acceptable outcomes. In Europe, the United States might like to see a unified Ukraine moving toward NATO and EU membership but might be able to accept a divided Ukraine with the larger part linked to Europe but forswearing NATO membership. In the Middle East, the United States would like to see a peaceful Syria under a moderate pro-Western government but, at this point, could probably live with a peaceful Syria under almost any government actually able to impose and maintain control. In East Asia, the United States faces little difficulty helping protect its treaty allies against actual invasion and occupation. The sources of friction with China tend, in some cases, to be issues in which the United States has no direct interest. Absent some major miscalculation, the risks for the United States and its allies are not existential. Some nontreaty partners and potential partners are more vulnerable, so there are legitimate cost-versus-risk equations to be evaluated when seeking to contain China.

This is not intended to suggest a race to the lowest acceptable outcome but, rather, to note that the cost-benefit ratio associated with achieving more than the minimum acceptable outcome needs to be part of the decision process. Declaratory policy will naturally set out desired outcomes; actual strategy must leave room for acceptable outcomes that fall short of declared goals. Strategic failure will ensue when the resources committed prove insufficient to reach even the minimally acceptable goals.

It is also worth remembering that there are disasters from which recovery may be impossible. Today, as during the Cold War, nuclear war with Russia—or even China—falls into this category. At some point, climate change might reach a point of no return, wherein no level of belated remedial action could reverse its progression. Accepting the risk of reversible damage is more justifiable than accepting risk of damage from which no recovery is possible.

Those who believe that the United States should spend less abroad and more at home, or that government should do less both at home and abroad, will be most willing to scale back both commitments and objectives and thus to accept, knowingly or not, the greater risk of failure and its attendant costs. At the other end of the spectrum, those who advocate aggressively for taking more action abroad will be the most willing to commit resources and the least willing to accept second-best outcomes, although they may underestimate the risks attendant on the pursuit of maximalist objectives.

Values and Interests

Given the range of threats it now faces, the United States cannot afford to define its purposes around principles as simple as containing Communism or combating terrorism. The challenges are too diverse, the battle lines too diffuse. China and Russia may be geopolitical competitors, but their cooperation will be needed to stem the proliferation of weapons of mass destruction, combat violent extremism, and deal with climate change. Iran is operating in parallel with the United States in Afghanistan and Iraq even as it opposes American purposes in Syria. China is a pillar of the global economy, a major trading partner, and the United States' largest creditor.

Even at the dawn of the Cold War, George Kennan, the father of containment, cautioned against looking at the world through a single prism. Reacting to what he regarded as President Harry Truman's over-the-top commitment in 1947 to support *any* nation threatened by communism, Kennan was

> struck by the congenital aversion of Americans to taking specific decision on specific problems, and by their persistent urge to seek universal formulae or doctrines in which to clothe and justify particular actions.

Containment, in Kennan's view, was a specific response to a specific problem, not an organizing principle for America's global engagement:

> Whatever the origins of this tendency, it is an unfortunate one. It confuses public understanding of international issues more than it clarifies it. It shackles and distorts the process of decision making. It causes decisions to be made on the basis of criteria only partially relevant or not relevant at all. It tends to exclude at many points the discrimination of judgment and the prudence of language required to the successful conduct of affairs of a great power.[3]

The next administration, like its predecessors, must present its national strategy as consistent with American values and interests. The ultimate test of leadership, after all, is the ability to inspire followers. Washington needs to enunciate policies that the public will support, partners will join, and adversaries will respect.

At a certain level of abstraction, this is easy enough to achieve. The United States values democracy and free markets and is interested

---

[3] George F. Kennan, *Memoirs 1925–1950*, New York: Pantheon, 1983, pp. 322–332.

in peace, international collaboration, and expanding trade. It is easier to collaborate with established democracies and to trade with free-market economies than to do so with authoritarian governments and closed economies. Therefore, U.S. values and interests cohere.

In reality, promoting values can be rather more complicated. Nondemocratic regimes resist and resent efforts to remake them in our image and will sometimes withhold collaboration on otherwise shared interests as a result. Freedom, democracy, and human rights may be universally applicable, but, much as we would wish otherwise, they are not universally attractive, particularly within conservative societies in which gender inequality and authoritarian rule are sanctioned and even enforced by religious authority. Finally, as the aftermath of the Arab Spring has demonstrated, there are worse things than a cooperative authoritarian government, such as an uncooperative and even more authoritarian government, as in today's Egypt, or anarchy and bloodshed, as in today's Yemen, Libya, and Syria.

In practice, therefore, there is sometimes tension between promoting democracy and human rights and advancing security and economic interests abroad. This requires case-by-case assessments of the local receptivity to such efforts; the cost of pressing too hard in terms of other issues; and the likelihood, if change comes, that it will move in the right direction.

Keeping Pace with Change

The world is more dangerous than it was a few years ago and the mounting chaos in the Middle East has fed wider anxieties. Many feel that the pace of technological change is quickening, that the international order is disintegrating, that power is shifting from national governments to individuals and nonstate actors, and that America's capacity to lead is waning. On balance, these portents of decline and disaster are overstated. Certainly, computers and the Internet are driving rapid change, but the pace is not more rapid or revolutionary than that following the introduction of electricity, radio, telephones, internal combustion engines, airplanes and the atomic bomb. The Middle East is in turmoil, but even taking account of the chaos in that region, inter- and intrastate conflicts continue to decline, as do the casualties and destruction these produce. States are being challenged by terrorists and insurgent groups in the Middle East, as they once were in Southeast Asia, Latin America, and the Balkans, but governments are actually more capable today in those once-turbulent regions, and they remain so in Europe, East Asia and North America. The Chinese economy has grown compared to that of the United States, but the United States has for many years been growing faster than Europe, Russia, and the rest of East Asia. Russia is misbehaving but not on the scale of the former Soviet Union.

In terms of shifts in global power balances, prior eras have been even more marked by rapid change. World War I ended with the collapse of the Austrian and Ottoman empires and the creation of a dozen new countries. In the two decades after World War II, control over more than one-half the world's surface and population shifted radically as dozens of nonstate actors—then known as liberation movements—seized power and set up new regimes. This occurred both in established states, such as China and Cuba, and throughout the colonial world. Even without the benefit of television and the Internet, the contagion of rebellion spread quickly from one society to another. Change occurred even more rapidly during the first two years of the George H. W. Bush administration, with the reunification of Germany, the disintegration of the Warsaw Pact, and the collapse of the Soviet Union. Again, there was broad contagion, with similar political changes occurring throughout Eastern Europe and 15 new states emerging from the former Soviet Union. Further, with the end of the Cold War came the quick settlement of a number of civil wars in Africa, Asia, and Latin America that had been stimulated by superpower competition.

These 20th-century power shifts exceeded anything we are experiencing today, but nearly all these changes were consistent with American interests and thus not occasion for grave concern—in contrast with current developments in the Arab world. Not surprisingly, people seem more struck by the pace of change for the worse than for the better. Whether recent events are to our liking or not, however, it would be hard to maintain that the distribution of power among states is changing more quickly today than it did in the years after 1918, 1945, or 1989.

This leaves the question of the power distribution between states, on the one hand, and nonstate actors, including individuals, on the other. Are states losing their grip? Is power devolving downward? An oft-heard concern is that the interstate system established by the Treaty of Westphalia in 1648 is unraveling. But most governments in Europe, East Asia, and the western hemisphere have not experienced diminished capacity. Africa has long been home to a number of failed and failing states, but the problems there are no more acute today than at any time since the decolonization of that continent some 60 years ago. What is new and disturbing is the fragility of Arab states. Several have descended into civil war. The rest are deathly afraid of doing so, leading their governments to take extreme and often ill-considered measures to counter what they regard as the forces of dissolution.

Regional upheavals of this sort are by no means a new phenomenon. In the 1950s, anticolonial revolts, generally employing terrorism, created dozens of new countries. In the 1960s and 1970s, all of Southeast Asia was engulfed in conflict. In the 1980s, Latin America and sub-Saharan Africa suffered multiple civil wars. At the end of the Cold War, the Balkans exploded. Throughout these decades, most Middle Eastern regimes exhibited a rigid resistance to change that has contributed to their current fragility.

Another concern is that even well-functioning states are losing ground as power is dispersed downward and outward. The continued expansion of international trade, finance, travel, and communications has increased vulnerabilities even as it has widened horizons, expanded opportunity, promoted economic growth, and extended longevity for billions of people around the world. Terrorists and criminals can mix with the millions of tourists and business people who cross international borders every day, but security agencies also have new and more powerful tools to track and impede their movements. Contagious disease can spread more rapidly, but resources to contain it can also be mobilized and dispatched more quickly. The communications revolution empowers individuals and states alike. Violent extremists can more easily spread their ideology, recruit followers, and orchestrate attacks, but security authorities can more easily collaborate to foil these attempts.

The Arab world aside, states are not fragmenting, nor are most of them losing ground to individuals or groups, malign or otherwise. Technology is neutral and can be used both to challenge and extend state authority. Admittedly, there is a race between the forces of order and disorder in all these domains, but it is not a race that states are predestined to lose. States have the authority and, in most cases, the capacity to regulate trade, finance, travel, and communications for the common good. They also have the power to employ these and other instruments to intimidate or damage other states, groups, and individuals. And, so far, only states have nuclear weapons. Thus, the greatest threats to international order continue to come not from terrorists, hackers, or other criminals, but from states—from strong states that threaten their neighbors and from weak states that cannot control outlaws in their midst.

Choices for America

The next eight chapters lay out major foreign and security policy choices that the next administration will face. Chapters Two through Six address issues of global relevance: the world economy, the U.S. defense budget and programs, counterterrorism (CT), cybersecurity, and climate change. The three following chapters are devoted to the regions where these challenges are most acute: Europe, East Asia, and the Middle East and South Asia.

In choosing issues to address, we have concentrated on those most likely to arise in the 2016 presidential campaign, which is to say those on which views are most likely to diverge. Rather than offering a single course of action, we identify plausible alternative courses that a responsible leader might choose. In some cases, these alternatives are clearly distinguished and mutually exclusive; in others, they represent illustrative points along a spectrum of alternatives. We recognize that, in practice, policymakers will often mix and match and that the resulting choices may in the end combine elements of the alternatives we offer.

The final chapter returns to the relationship between policy and strategy, proposing the ends, ways, and means that should guide an American national strategy for the 21st century.

CHAPTER TWO

The Global Economy

The acceleration of U.S. gross domestic product (GDP) in mid-2014 raised hopes that the United States might have finally returned to sustained economic growth after the Great Recession, which officially started in December 2007 and ended in June 2009 (Figure 2.1). But since then, signals have been mixed. Growth was lower than expected in the fourth quarter of 2014, and negative in the first quarter of 2015. In addition, employment increases were lower than expected in early 2015 and, even with increases of 254,000 and 223,000 in May and June 2015, the first-half 2015 monthly average registered well below the first-half 2014 monthly average.

Nevertheless, the International Monetary Fund (IMF) deems renewed U.S. growth likely. In its October 2014 *World Economic Outlook*, the IMF projected U.S. 2015 GDP growth at 3.1 percent, fifth-highest among 35 advanced economies of all sizes.[1] In its January 2015 update, the IMF actually raised the projection for U.S. growth in 2015 to 3.6 percent, while lowering those of the euro area, Japan, Canada, and all the so-called BRICS—Brazil, Russia, India, China, and South Africa.[2] The same holds for the 2016 projections: the IMF raised its projection of U.S. growth by 0.3 percentage points to 3.3 percent, while lowering its projections for the euro area, Japan, Canada, and all of the BRICS except India.

---

1 IMF, *World Economic Outlook: Legacies, Clouds, Uncertainties*, World Economic and Financial Surveys, Washington, D.C., October 2014.

2 IMF, *World Economic Outlook Update*, Washington, D.C., January 2015a.

Figure 2.1
Growth of U.S. Gross Domestic Product

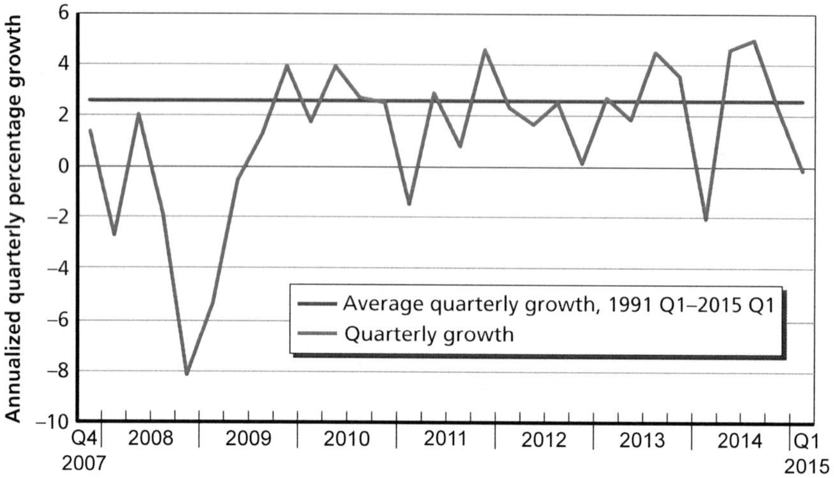

SOURCE: Bureau of Economic Analysis, "Percent Change from Preceding Period," spreadsheet, U.S. Department of Commerce, May 29, 2015.
RAND *RR1114-2.1*

For some, the Great Recession called into question the value of the U.S. economic model. Combined with continued robust growth in China and the developing world during the postrecession period, the recession raised questions about the ability of the United States to lead the global economy as it had done since the end of the Second World War. Therefore, in considering the economic component of America's role in the world, U.S. policymakers will need to consider not only what the United States wants but what it has the power to bring about. For now at least, that power is considerable.

Trends and Challenges

Although the United States will hold its place as the world's largest economy for many years in the future, the environment in which it is operating is changing. The global economy is far more integrated now than even just 25 years ago, and advancements in communications and technology have both accelerated the spread of information and increased its volume. In addition, even though the U.S. economy is sizable and growing, the U.S. government faces a sizable and growing debt burden that may become unsustainable and must be dealt with before it becomes so.

Gross Domestic Product and the U.S. Share of the World Economy

The United States remains the world's largest economy. Although the U.S. share of the global economy declined by 3.9 percentage points between 1991 and 2013 (comparable 2014 data are not yet available as of this writing), the U.S. economy still constitutes more than one-fifth of the global economy (Figure 2.2). In addition, much of this decline took place after the financial crisis; if U.S. growth continues, the Chinese slowdown continues, and declines in Europe and elsewhere continue—as the IMF projects—this recent decline may reverse.

A greater concern than the relative trend line of U.S. GDP is that of U.S. allies; they are performing far more poorly. In particular, in 1991, the EU's economy constituted 33.2 percent of nominal global GDP,

Figure 2.2
Share of the Global Economy

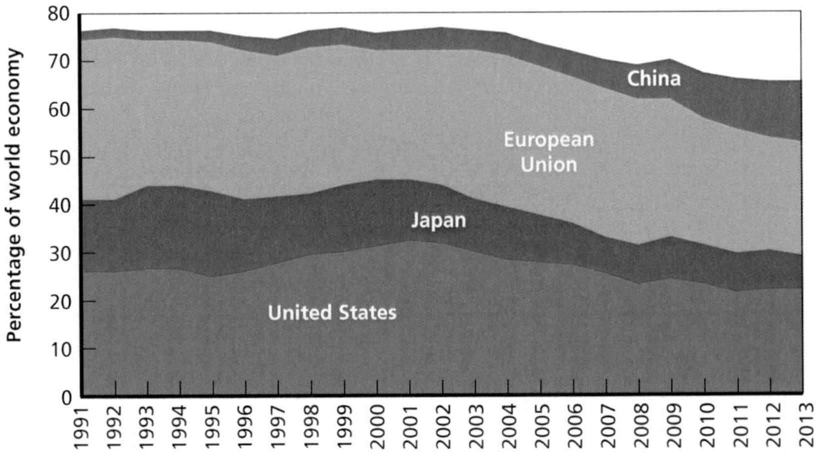

SOURCE: The World Bank, "World Development Indicators," web page, April 14, 2015.
RAND *RR1114-2.2*

but by 2013, that figure was 23.8 percent. Japan's share of the nominal global economy has also declined a great deal, although it was much smaller to begin with: In 1991, it totaled 15.0 percent, but by 2013, that had fallen by more than half. On the other hand, the global economy remains heavily dominated by free-market democracies despite the oft-cited appeal of alternative economic and political models.

One other important trend is the broadening of economic growth. In 1991, the economies of the United States, the EU, Japan, and China constituted 75.8 percent of global GDP. But by 2013, this share had fallen to 64.7 percent. Many other economies grew more rapidly as well. This is likely to continue. In its January 2015 *World Economic Outlook* update, the IMF projects year-on-year growth in the advanced economies to be 2.4 percent in both 2015 and 2016 but growth in emerging markets and developing economies to be 4.3 percent in 2015 and 4.7 percent in 2016. Among the emerging markets and developing economies, much of that growth is expected to come from China, India, the Association of Southeast Asian Nations, and sub-Saharan Africa. Growth in Latin America and the Caribbean is projected to lag behind even that of the advanced economies.

The United States will remain the world's largest national economy likely through at least two more decades. This supposes no rapid deceleration of or crisis in the Chinese economy and the United States maintaining a growth rate close to its historical average. But if Europe and Japan continue to struggle, the United States and its allies as a group will likely lose weight in the global economy. Still, as Figure 2.2 suggests, market democracies will continue to dominate the international economy, and the advanced countries will continue to be the most desirable markets because of their size and sophistication.

Economic Interdependence

The United States played the leading role setting up the rules-based international economic order that enabled a dramatic expansion of trade and investment, especially since the end of the Soviet Union in 1991. That year, the value of exports from all countries was equal to 19.1 percent of global GDP. That figure hit a high of 31.8 percent in 2008 and was hardly dented by the Great Recession, measuring 31.0 percent in 2013.

Foreign direct investment (FDI) has grown even more rapidly. FDI is cross-border investment to control an operating business or real estate and is another measure of economic interdependence. Components of FDI include equity investment, loans by a parent company in one country to its subsidiary or branch in another country, and earnings retained by the subsidiary or branch and not sent back to the parent company. In 1991, the stock of FDI from all countries relative to global GDP was 9.9 percent. By 2013, that figure was 33.7 percent, growing more than three times in relative terms.

Communications and Technology

Not only is the world more integrated in terms of goods and services, it is more tightly bound in terms of information, both the volume of information flowing and the speed with which it flows. This means reaction times for market participants and policymakers are often quicker than in the past. As one sign of this global information revolution, worldwide mobile cellular subscriptions per 100 people grew from 0.3 in 1991 to 92.6 in 2013. By 2013, there were almost as many cellular subscriptions as there were people on the planet.

Fueling this growth has been the growth of exports in information and communications technology (ICT) goods and services. As one sign of this, exports of ICT services as a percentage of all services exports grew from 27.6 percent in 2005 (the earliest year for available data) to 32.8 percent in 2013, nearly one-third of all services exports.

The United States and its major allies have so far been in the lead on the ICT revolution, in part through high levels of innovation. The United States has the largest global share of knowledge-intensive service industries, at 32 percent of the global total, and the largest share of high-technology manufacturing, at 27 percent, although China is a close second in the latter category.[3] In terms of exports, the United States is second behind the EU in knowledge-intensive services, and

---

[3] National Science Board, *Science and Engineering Indicators 2014*, Arlington, Va.: National Science Foundation, NSB 14-01, 2014.

third behind China and the EU in manufactured high-technology products.

Difficult Budget Choices

Alongside this generally positive assessment of the U.S. position in a changing global environment, the United States faces potentially severe long-term budget challenges. Spurred by the financial crisis, federal debt held by the public, both U.S. and foreign, has more than doubled relative to the size of the economy, to an estimated 74 percent in 2014. Although this debt was almost 50 percent of GDP in the early 1990s, it fell to as low as 31 percent in 2001 and had risen to only 35 percent by 2007, the year of the onset of the recession preceding the financial crisis.[4]

Some of this is to be expected. During recessions, so-called automatic stabilizers, such as unemployment benefits, rise. This helps people smooth their consumption during hard times. A major federal stimulus program also swelled the debt.

However, even if both these phenomena prove temporary, the nation faces severe longer-term problems. These are driven largely by increases in retirement payments and health care costs as the population ages. The result is ever-increasing deficits and ever-increasing debt. If current law continues, the deficit is expected to rise from 2.8 percent of GDP in 2014 to 6.5 percent of GDP by 2040, and the debt is expected to rise from 74 percent of GDP to 108 percent of GDP by 2040 (Figure 2.3).[5]

While spending on Social Security, Medicare, Medicaid, the federal Children's Health Insurance Program, and health insurance subsidies is expected to be 55.7 percent of federal revenues in 2014, it is expected to reach 87.4 percent of federal revenue in 2089, leaving little money available for much else. In successive remarks from 2010 through 2014, then–Chairman of the Joint Chiefs of Staff Admiral

---

[4] Congressional Budget Office, *The 2014 Long-Term Budget Outlook*, Washington, D.C., July 2014a.

[5] Over the even longer term, the deficit is expected to rise to 12.7 percent of GDP, and the debt is expected to rise to 225 percent of GDP by 2089. Congressional Budget Office, 2014a.

Figure 2.3
Projected U.S. Federal Deficit and Debt

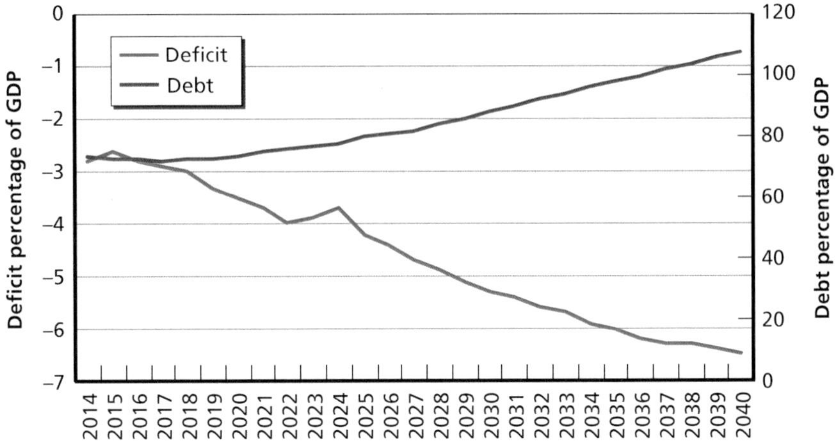

SOURCE: Congressional Budget Office, 2014a.
RAND *RR1114-2.3*

Mike Mullen called the debt the single biggest threat to U.S. national security.

The United States will struggle in the short run with its budget policy. Fixing the long-term budget problems will have to include some combination of budget cuts and revenue increases. Until this bargain is consummated, resources for defense and diplomacy will dip below historic levels, or the national debt will grow even more rapidly.

Channeling Chinese Growth

The U.S.-China relationship is likely to prove pivotal in U.S. economic and security policy. China's economic interests have become global, and its security interests are expanding. Its economic relationship with the United States has also been expanding. In 2011, the year of latest available data, U.S. goods and services exports to China, plus sales in China by U.S.-owned affiliate companies, totaled $278 billion, or 5.2 percent of all U.S. exports plus sales by U.S.-owned affiliates abroad. U.S. goods and services imports from China, plus total sales by Chinese-owned companies established in the United States, totaled

$422 billion, or 6.6 percent of all U.S. imports plus sales in the United States by foreign-owned companies in the United States.[6] These figures do not include the substantial sales to China by U.S.-owned companies located elsewhere in the region.

Despite its rapid growth, China faces a number of internal weaknesses that will make it hard to sustain this growth. China's economy is uniquely unbalanced, with levels of investment as a share of GDP that are far above those of any rapidly growing East Asian country, even when the countries were at China's level of development. This reflects inefficient overinvestment.[7] The flipside is that China has had levels of consumption as a share of GDP that are far below those of any rapidly growing East Asian country. Furthermore, Chinese workers do not receive a large share of the income generated in their economy, known as the labor share of the economy. With such a small labor share, household consumption has been unusually low.

Chinese officials are well aware of these issues. Without a new growth model, China is unlikely to keep growing rapidly.[8] Accordingly, following the Third Plenary Session of the Chinese Communist Party Central Committee in 2013, the Chinese authorities issued an ambitious document with 60 reform steps.[9] Although it is not yet possible to tell whether these reforms are being implemented, China recently signed an agreement with the IMF aimed at strengthening

---

[6] The two aggregates are not entirely comparable, since data are not available for sales by U.S.-based Chinese affiliates only to the United States, rather than total sales, whereas data are available on sales specifically to China by U.S.-owned companies in China. If we used total sales by U.S.-owned companies in China, rather than just sales to China, then U.S. goods and services exports to China plus sales by U.S.-owned companies there would total $329 billion, or 4.3 percent of all U.S. goods and services exports and foreign-affiliate sales.

[7] David Dollar, "China's Rebalancing: Lessons from East Asian Economic History," Working Paper Series, Washington, D.C.: John L. Thornton China Center, Brookings Institution, October 2013.

[8] Richard Cooper, "The Third Plenum and Economic Reform," *Harvard Economics Review*, Spring 2014, pp. 12–14.

[9] "The Decision on Major Issues Concerning Comprehensively Deepening Reforms in Brief," *China Daily*, November 16, 2013.

fiscal institutions, reflecting a number of reforms called for in the 2013 document.[10]

Whatever China's growth rate, its involvement in international institutions will be relevant to the next administration. First, the United States and its partners must determine whether and how to integrate China into the Trans-Pacific Partnership (TPP). China is not now among the negotiating partners and will not be until the treaty is concluded. Second, the United States must determine its stance toward the China-inspired New Development Bank and Asian Infrastructure Investment Bank (AIIB). Finally, the United States is negotiating a bilateral investment treaty (BIT) with China. Such a treaty could lead to a much greater opening of China for U.S. investment and better treatment of U.S. investors in China. Additionally, it would include a dispute-settlement mechanism that moves away from the potential biases of national courts. The policy challenge for the United States is to balance the benefits of completing such an agreement with the rigor of the treaty and how it compares with other BITs the United States has negotiated and plans to negotiate.

In the sections that follow, we present two strategic choices U.S. policymakers face that will determine the U.S. role in the world. In both cases, we present two sets of policy options. It is likely that the most effective policy choices will come from a blending of these sets.

Including or Excluding China

In the realm of international finance, the IMF and World Bank, founded by the United States and the allied powers after World War II, are still dominated by those countries. Other countries have grown and become larger contributors to these institutions and to the health of the global economy without commensurate gains in their power over

---

[10] IMF, "IMF and China's Ministry of Finance Sign Agreement on Strengthening Fiscal Institutions and Capacity Development," press release, Washington, D.C., No. 15/141, March 27, 2015b.

governance. Some, including China, have become global lenders and donors as well.[11]

In 2010, both the World Bank and the IMF introduced major reforms in the voting power held by their member countries. Despite the changes, the United States remains the only single country that can veto major governance changes. The U.S. Congress has not yet approved certain of these IMF governance changes, so they have not gone into force. Partly because of this but certainly for other reasons, major developing countries—led by China—have embarked on establishing new multilateral institutions, namely the New Development Bank and the AIIB. Both institutions have been expected to put developing countries, rather than the developed countries, in charge. Analysts indicate that China and its partners could achieve a number of objectives through these institutions, certainly increasing China's influence, but also diluting the power of existing international institutions by avoiding some of the rules they impose.

The AIIB will be a particularly important test case, as the United States has lobbied its allies to stay out of the institution. Until January 2015, Singapore was the only non–oil-exporting economically advanced country to sign up as a founding member. New Zealand joined in January, and then, in March, numerous U.S. allies decided to join, including the United Kingdom, France, Germany, and Italy. They argued that they can best shape the institution from the inside, to make sure it meets the standards of the existing major development banks, including the World Band and the Asian Development Bank. AIIB success could magnify China's role, but advanced-country participation could also mute China's influence. Until there is more of a track record, it is unclear whether the AIIB will advance Chinese power.

Bringing China In

U.S. policies toward multinational institutions will be linked to China policy. For example, the United States could fully support the Regional

---

[11] Arvind Subramanian, "Preserving the Open Global Economic System: A Strategic Blueprint for China and the United States," policy brief, Washington, D.C.: Peterson Institute for International Economics, PB13-16, June 2013.

Comprehensive Economic Partnership, the New Development Bank, and the AIIB and could approve reforms that would increase China's voting power in the World Bank and IMF. Other policies more specific to China could involve completing a BIT that is not quite as strong as U.S. BITs with other countries. The advantages would be to lock in an agreement that improves the climate for U.S. investors and to set the stage for further negotiations when both countries find such negotiations in their interest. Another accommodative policy could be working with China to help internationalize the Chinese currency, the renminbi. As of now, it has limited international use, but China is trying to change that. The United States could assist by including renminbi securities in U.S. international reserves and working with China to develop a renminbi market in the United States.

Keeping China Out

Alternatively, the United States could take a tougher stance toward China, with the aim of limiting its economic influence or channeling its development within U.S. standards. Such a policy stance would include continuing to limit China's voting power in multilateral institutions, delaying China's entry into the TPP until all TPP requirements are fully met, actively discouraging countries from joining the New Development Bank, and actively countering the AIIB and highlighting any problems it has or causes. It could also involve excluding the renminbi from U.S. reserves, working against its international adoption, and insisting on a BIT with standards equivalent to or better than the most advanced U.S. BITs. Note that agreeing to a high-standard BIT might actually be desired by Chinese reformers, since they could use such an agreement to drive their own agenda in China.

Global Versus Regional Rule-Setting

The rules-based international economic system established following World War II has rested on two main pillars. The first is multilateral trade liberalization through what was originally the General Agreement on Tariffs and Trade and what is now the World Trade Organization (WTO). The second is assistance through multilateral institutions for balance-of-payments adjustments and other financial and budget crises, under the IMF, and economic development, under the World Bank and regional development banks. The United States has been the dominant player in most of these institutions. Both pillars are under modest challenge. With economic growth broadening and the major economies' share of total world GDP shrinking, there is a strong case that a rules-based system is even more important because multiple players may have divergent interests and because actions taken by one for short-term advantage would harm the entire system over the longer term. With greater weight in the global economy comes less ability to free-ride at little cost to others.

Since the WTO entered into force in 1995, there has been no further global multilateral trade liberalization. The WTO talks started in 2001 under the rubric of the Doha Development Agenda have stalled, and nations are seeking other trade arrangements to increase economic integration. These other trade arrangements are much larger than previous regional agreements and have become known as mega-FTAs (free-trade agreements) or megaregionals.[12] The four largest are the TPP, involving 12 countries, including the United States, 38 percent of world GDP, and 24 percent of world exports; the Trans-Atlantic Trade and Investment Partnership (TTIP), involving 29 countries, including the United States, 46 percent of world GDP, and 25 percent of world exports; the Regional Comprehensive Economic Partnership, involving 16 countries, not including the United States, 29 percent of world GDP, and 30 percent of world exports; and the EU-Japan FTA, involving 29 countries, not including the United States, and more than one-third of world GDP.[13]

There is widespread agreement that such treaties lead to higher overall income for the parties, but critics fault trade and FTAs for pro-

---

[12] Ruben Van den Hengel, "The Rise of the Mega-FTAs," fact sheet, Singapore: EU Centre, October 2013; Jeffrey J. Schott, "Can Mega-Regionals Support Multilateralism?" presented at the OECD Global Forum on Trade, Reconciling Regionalism and Multilateralism in a Post-Bali World, Paris, February 11, 2014.

[13] Schott, 2014; European Commission, "Countries and Regions: Japan," web page, Directorate-General for Trade, September 15, 2014.

moting overseas outsourcing, widening U.S. income disparity, and spurring deindustrialization. The TPP and TTIP are also noted for having potential strategic benefits—the former is seen as validation of the rebalance to Asia and the latter as a countervailing gesture reinforcing the U.S. commitment to Europe.[14] In combination, conclusion of these agreements will consolidate the U.S. position as the lynchpin of the world's two largest trading blocks. Additionally, these treaties could serve as templates for future trade agreements.[15]

Regionalism

In the realm of multilateral trade rules, one set of options would be to continue to focus on regions where the United States could be the most successful in negotiations and where it could gain the most. After completion of TPP and TTIP, this would involve broadening TPP with a rigorous entry procedure and working to broaden TTIP to non-EU European countries. It would not involve trying to restart current WTO talks but could involve a U.S.–Middle East FTA to link together the Middle Eastern countries with which the United States already has bilateral agreements, along with other interested and capable countries in the region. In the realm of international institutions, policy would include maintaining the sole U.S. veto in the World Bank and IMF and ensuring that a coalition of countries, except for those in Europe and Japan, would have difficulty forming a veto coalition. The maintenance of U.S. institutional primacy would also include active opposition to the New Development Bank, trying to influence countries not to join, and trying to thwart the success of the AIIB. Despite such U.S. opposition, one effect of this option could be an expansion of parallel trade and financial institutions over which the United States would have little influence.

---

[14] Shayerah Ilias Akhtar and Vivian C. Jones, *Proposed Transatlantic Trade and Investment Partnership (T-TIP): In Brief*, Washington, D.C.: Congressional Research Service, June 11, 2014.

[15] Ian F. Fergusson, William H. Cooper, Remy Jurenas, and Brock R. Williams, *The Trans-Pacific Partnership (TPP) Negotiations and Issues for Congress*, Washington, D.C.: Congressional Research Service, December 13, 2013.

Internationalism

After completion of TPP and TTIP, the United States could attempt to restart global trade and investment talks, either by starting a process to merge TPP and TTIP or by restarting WTO talks. Renewed and aggressive internationalism could also involve creating a fast track for China, as one of the world's dominant trading powers, to join TPP and creating well-defined tracks for other Asian countries to do so. It could also involve embarking on talks to merge TPP with an Asian FTA now under negotiation that excludes the United States, the Regional Comprehensive Economic Partnership. In the realm of multilateral institutions, policy could including elevating the voting share of China and other developing countries in the World Bank and IMF commensurate with their contributions, which would allow them to more easily form a veto coalition. It would also involve the United States supporting both the New Development Bank and AIIB and seeking to become a member of both.

CHAPTER THREE

National Defense

The American strategy and defense program as put forward by the Obama administration in 2014 is predicated on the assumptions that Europe would be stable and at peace and that conditions in Iraq, Afghanistan, and the Middle East more broadly were such that large-scale U.S. ground and air force commitments could continue being drawn down, transitioning to a posture that would support a series of small-scale counterterror efforts. These conditions, it was thought, would allow the United States to focus more attention and resources on shoring up U.S. influence in the dynamic Asia-Pacific region—what the administration has called the "rebalance" to the Pacific. As has already been noted, those assumptions have been upended by Russia's military aggression against Ukraine and the collapse of Iraqi military forces in the face of attacks by the Sunni jihadist group ISIS.

Less obvious but nearly as important is the fact that the administration's current defense program assumes that Congress will, over the next five years, appropriate the funds requested in the President's fiscal year (FY) 2015 budget submission. That submission calls for defense spending through FY 2019 that totals $115 billion more than would be allowed by the Budget Control Act (BCA) of 2013, which has a cap of $499 billion for FY 2016.[1] Yet the BCA remains law unless and until it is overturned.

Further, while the defense strategy still calls for a force with sufficient capacity to fight and win in more than one region at a time, it

---

[1] Amy Belasco, *Defense Spending and the Budget Control Act Limits*, Washington, D.C.: Congressional Research Service, R44039, May 19, 2015.

does not contemplate the substantial day-to-day demands that will fall on U.S. forces in the three critical regions of Europe, the Middle East, and East Asia. Nor does the strategy explicitly account for several critical tasks, such as eliminating nuclear, chemical, and biological weapons in North Korea should the regime collapse. Taken together, these developments mean that U.S. defense strategy today is out of alignment with the resources that the nation has been expecting to devote to the defense program.

Indeed, even if the administration gets the extra $115 billion it has requested above the BCA caps for FYs 2015–2019, the resulting program would likely be insufficient to meet the demands of a world in which Russia is actively challenging the security and territorial integrity of neighboring states and in which U.S. air and ground forces remain engaged in Iraq, Afghanistan, and perhaps elsewhere in numbers that had not been anticipated one year ago. This imbalance between requirements and resources is further exacerbated by the fact that Congress has been consistently reluctant to approve a range of proposals to reduce the costs of Department of Defense (DoD) administrative, overhead, infrastructure, and personnel accounts. Since 2011, DoD has submitted budget requests that have called for modest reductions in the rate of growth of military pay, increases in copayments for military family health care, cuts to the subsidies provided to military commissaries, closures of unneeded bases, and other measures. Most of these proposals have been reversed by Congress. Estimates of possible savings from these sorts of measures range upwards of $20 billion per year.

Europe

Nowhere is the gap between U.S. security commitments and regional posture more pronounced than in Europe. In September 2014, in a high-profile speech in Tallinn, Estonia, President Obama underscored the U.S. commitment to defend its NATO allies with these words:

> [W]e will defend the territorial integrity of every single [NATO] ally. . . . Article 5 is crystal clear. An attack on one is an attack on

> all. So, if . . . you ever ask again, who'll come to help, you'll know the answer: the NATO alliance, including the armed forces of the United States of America . . . You lost your independence once before. With NATO, you will never lose it again.[2]

The President's speech was made the day before the NATO Summit in Wales—a meeting that took place while Russian forces were actively prosecuting military operations against Ukraine. In response to that aggression, NATO's member states have taken steps and made plans to strengthen the Alliance's defensive posture and shore up deterrence. Among these steps are the following measures:

- persistent deployment of U.S. ground forces (generally battalion- or company-sized entities) to the Baltics
- creation of 40-person multinational NATO force integration units in all frontline states
- stepped-up levels of air policing activity in the Baltics
- U.S. training for Polish F-16 pilots, including periodic U.S. F-16 deployments
- creation of a very high readiness joint task force—a brigade-sized, multinational unit that would be prepared to deploy across the NATO treaty area within six to ten days
- prepositioning of military stocks in frontline countries
- pledges (so far unfulfilled) to increase European defense capabilities.

These and other measures should have salutary effects on the military balance on the Alliance's eastern flanks. However, even if fully implemented, the resulting NATO posture in areas contiguous to Russia and Belarus would not support a credible defense against a determined Russian attack. In summer 2014, as President Vladimir Putin sought to coerce the leaders of Ukraine into accepting a de facto Russian-dominated state carved out of its eastern provinces, he was able to muster as many as 40,000 troops across the border. Those units con-

---

[2] Barack Obama, "Remarks by President Obama to the People of Estonia," Washington, D.C.: The White House, Office of the Press Secretary, September 3, 2014b.

sisted of armor, artillery, mechanized and heliborne infantry, special operations, and tactical aviation forces and were judged to be combat ready. Given the geography of and transportation networks in western Russia, forces of this magnitude can be deployed to border regions within days to at most weeks, while NATO today would need months to deploy a comparable force to its eastern flanks.

The gold standard of deterrence and assurance is a defensive posture that confronts the adversary with the prospect of operational failure as the consequence of aggression. While analysis of potential scenarios involving Russian aggression in this part of the world has only recently begun, it is quite clear that, in many plausible scenarios, NATO forces today would be unable to defeat or even meaningfully impede a sizable armored invasion of one or more of the Baltic states. NATO air forces—flying from bases in Germany, the United Kingdom, and other allied nations, and quickly reinforced from outside the area—could bring significant firepower to bear against such an invasion. However, it would take time to suppress Russian air defenses protecting the attacking ground force and—without a strong NATO ground force to compel the attackers to slow their movement and mass—airpower would be unlikely to have decisive effects.[3]

Three to four heavy brigades, totaling approximately 15,000 to 20,000 troops, would be sufficient to deny Russia the prospect of a *coup de main* from a nonalerted posture and, as such, could greatly strengthen deterrence.[4] The United States could provide two of these brigades, with the remainder being provided by NATO allies. These forces need not be permanently deployed in the Baltic states, but most of their heavy equipment,[5] consumables (e.g., ammunition, fuel, spare

---

[3] NATO could, of course, plan to subsequently retake the disputed areas—at some cost—but such a posture would do little to assure the NATO member states that felt threatened, and Russia might use or threaten to use nuclear weapons to defend its gains. This underscores the importance of preventing Russia from seizing allied territory in the first place.

[4] As noted previously, analysis of these scenarios is in an early phase, and these estimates reflect preliminary findings only. But we can say with confidence that moving now toward a posture of this scale would be an appropriate step toward a viable deterrent posture.

[5] A recent proposal from the Pentagon suggests that U.S. heavy equipment and weaponry will soon be prepositioned in several Baltic and Eastern European states. See Eric Schmitt

parts), and supporting elements (e.g., logistics, communications gear, mobile air defenses) should be prepositioned forward, with a continuous presence of battalion-sized rotational forces. Follow-on NATO ground forces would be required to reinforce this initial defending force and to provide a stalwart defense against a mobilized Russian attack.

Other steps to enable effective defensive operations could include prepositioning air-delivered antiarmor munitions at bases in Germany and Poland and ensuring that NATO air forces have capabilities and concepts for rapid suppression of the enemy's integrated air defenses and for cruise missile defense. NATO forces would also need to constitute a deployable headquarters to command multinational ground operations from a forward location.

East Asia

The growth of China's military power poses serious threats to the viability of the United States' role as the security partner of choice for Japan, South Korea, Australia, the Philippines, Taiwan, and other states in the region. Chinese military writings are replete with discussions of how to fight a "local war under high-technology conditions" against a technologically superior foe, such as the United States. Chinese strategists have carefully studied recent U.S. military operations in efforts to identify U.S. vulnerabilities and have devised strategies to exploit them. Such strategies include attacking air bases, ports, and aircraft carriers; information systems, such as sensors and communication nodes, including satellites; and logistics assets, including supply depots and naval replenishment ships.[6] China's armed forces are rapidly acquiring the wherewithal to undertake such attacks.

---

and Steven Lee Myers, "U.S. Is Poised to Put Heavy Weaponry in Eastern Europe," *New York Times*, June 13, 2015.

6 Roger Cliff, Mark Burles, Michael S. Chase, Derek Keaton, and Kevin L. Pollpeter, *Entering the Dragon's Lair: Chinese Antiaccess Strategies and Their Implications for the United States*, Santa Monica, Calif.: RAND Corporation, MG-524-AF, 2007, pp. xv–xvii.

Unlike in Europe, the problem with the U.S. defense posture in Asia is not primarily one of inadequate numbers of U.S. forces deployed forward. Today, the United States deploys approximately 325,000 servicemen and women in U.S. Pacific Command, with forward-deployed forces that include a Navy carrier strike group home ported in Japan; eight Air Force and Marine Corps fighter squadrons; 12 attack submarines and one to two cruise missile submarines; one Marine amphibious ready group; and periodic rotations of fighter, bomber, and tanker aircraft to Guam.[7] In a crisis, these elements could be reinforced by rapidly deploying air and naval forces. The problem is that U.S. forces in the region—particularly land-based air forces, fixed infrastructure ashore, and naval surface vessels—are vulnerable to attack by Chinese long-range precision strike assets, principally cruise and ballistic missiles. In a crisis, this risks creating a situation in which U.S. efforts to strengthen deterrence and stabilize the situation by sending more forces to the region could actually have the opposite effect, provoking China's leaders into attacking lucrative targets as a means of gaining the initiative in a conflict—what some call a first-mover advantage.

If U.S., allied, and partner forces are to retain credible capabilities to deter and defeat an adversary with advanced military capabilities, new investments in platforms, weapons, infrastructure, and support systems are needed. But meeting the challenge will require more than simply buying and fielding new and better gear. The magnitude of the antiaccess and area denial (A2/AD) capabilities posed by Chinese long-range strike and other weapons is such that new concepts for the conduct of power projection operations are needed. Money, time, and talent must therefore be allocated not only to the development and procurement of new equipment and infrastructure but also to concept development, gaming and analysis, field experimentation, and exploratory joint force exercises.

Perhaps most urgent is the need for new approaches to basing and operating forward forces in the A2/AD environment. Meeting this

---

[7] International Institute for Strategic Studies, "North America," *The Military Balance*, London: Routledge, 2014; Michael J. Green, Gregory Kiley, Nicholas Szechenyi, and David J. Berteau, *U.S. Force Posture Strategy in the Asia Pacific Region: An Independent Assessment*, Washington, D.C.: Center for Strategic and International Studies, 2013, p. 48.

challenge will involve a mix of approaches, including selective hardening of key facilities at bases and enhancing the ability of engineering teams at these locations to rapidly repair damaged infrastructure. Such steps must be complemented by efforts to ensure that U.S. air forces in the western Pacific are able to operate from dispersed bases, including austere facilities with little in the way of base infrastructure.

Another key to defeating sophisticated A2/AD threats will be for U.S. and allied forces to develop better reconnaissance and strike capabilities of their own to be able to destroy the enemy's attacking forces early in a conflict. Over the past 25 years, U.S. forces have become accustomed to dominating all five domains of military operations—air, land, sea, space, and cyberspace—virtually from the outset of a large-scale operation. For this reason, U.S. forces have not, for the most part, invested in capabilities for reconnaissance and long-range strike in contested environments. Developing and deploying penetrating platforms and standoff weapons, such as cruise missiles, in wider varieties and larger numbers could help to change this.

Counterterrorist Operations

Since 9/11, the terrorist threat to the United States and its interests abroad has metastasized from the fairly hierarchical structure of Osama Bin Laden's al Qaeda to an increasingly decentralized threat comprising al Qaeda, with its affiliates and imitators; a panoply of other Salafi-jihadist groups, such as ISIS; Hezbollah, which remains active in such areas as the Middle East, Africa, and Latin America; and radicalized individuals and networks at home and abroad. Trends are not positive, as jihadist battlefields in such countries as Syria and Iraq will likely continue to serve as training grounds for foreign fighters, including some Americans and other Westerners.

Figure 3.1 provides a rough estimate of the number of Salafi-jihadist fighters between 1988 and 2013. Calculating the number of Salafi-jihadists is difficult, in part because such groups do not provide public estimates of their numbers, which can vary considerably over the course of a group's life. Consequently, Figure 3.1 depicts high and

Figure 3.1
Estimated Number of Salafi-Jihadists, by Year, 1988–2013

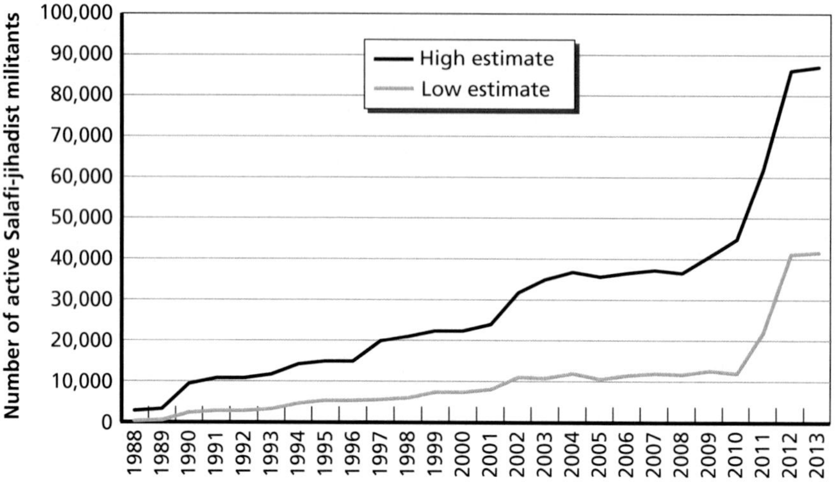

SOURCE: Seth G. Jones, *A Persistent Threat: The Evolution of al Qa'ida and Other Salafi Jihadists*, Santa Monica, Calif.: RAND Corporation, RR-637-OSD, 2014, p. 27.
RAND *RR1114-3.1*

low estimates for the number of Salafi-jihadists by year. The sharp increase in the number of fighters after 2010 is mostly attributable to the conflicts in Syria and Iraq.

Figure 3.2 depicts the number of attacks by core al Qaeda and its affiliates for each year since 2007. The data show that violence levels are highest in Yemen (from al Qaeda in the Arabian Peninsula), Somalia (from al Shabaab), Iraq (from ISIS), and Syria (from Jabhat al-Nusrah and ISIS). Approximately 98 percent of al Qaeda's and its affiliated organizations' attacks over that seven-year span were against "near enemy" targets (opponents in the country or region where the group is headquartered) and only 2 percent were against "far enemy" targets (opponents outside the region, primarily in Europe or the United States). The trends in numbers of casualties and fatalities inflicted by these groups are similar.

The persistence of terrorist threats is attributable to two major factors in the international system: the weakness of governments across Africa, the Middle East, and South Asia, which creates opportunities

Figure 3.2
Number of Attacks by Al Qaeda and Affiliates, 2007–2013

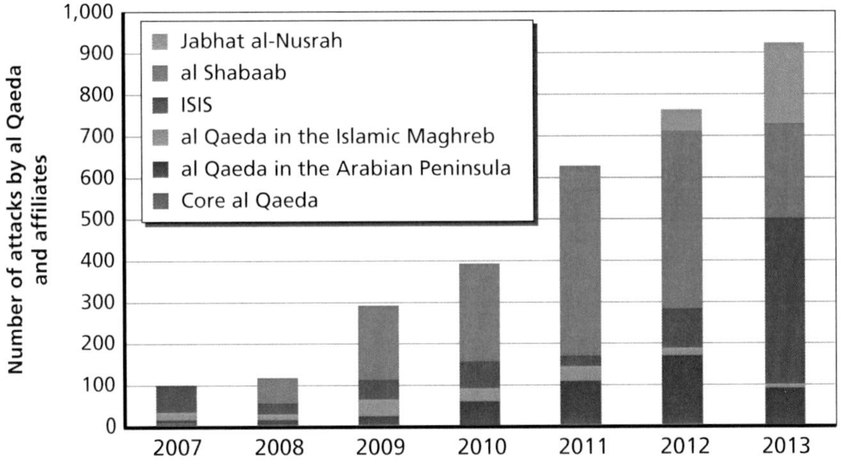

SOURCE: Jones, 2014, p. 35.
RAND *RR1114-3.2*

for terrorist groups to gestate and operate, and the transnational spread of militant networks. Operatives who spend time training or fighting in such countries as Iraq, Afghanistan, Pakistan, and Syria have been able to move into North Africa, the Levant, and even to Brussels and Paris. As mastery over more-destructive technologies continues to devolve to lower levels of human organization, from nation-states to subnational groups to individuals, terrorist groups will become potentially more potent over time. The January 2015 attack against the French satirical magazine *Charlie Hebdo* in Paris also suggests that terrorist groups can execute attacks with simple weapons, such as assault rifles. Al Qaeda in the Arabian Peninsula provided training to two of the operatives involved in the shootings, Said and Chérif Kouachi.

All this means that the United States, along with its allies, may need to conduct a campaign against Salafi-jihadist groups for the indefinite future, both overseas and at home. With the movement of foreign fighters to and from the West, there is a growing need to stop the flow with improved intelligence collection and sharing, border interdiction efforts, and legal measures. The specific loci of that campaign will shift over the years, with CT assets deploying to combat the most acute threats. U.S. CT forces will engage in two broad types of operations: indirect approaches, in which U.S. forces work to help build the capacity of partner security forces, providing training, advice, and assistance, and direct action, in which forces conduct precision attacks on terrorist groups and their financial, logistical, and political support networks.

Since 2008, the number of U.S. special operations forces (SOF), including military and civilian personnel, has grown from 54,200 to approximately 70,000. DoD should plan to sustain at least this level of SOF indefinitely. It may also want to increase funding for SOF training and equipment. Top priorities for enhanced training include basic and advanced special operations skills and increased foreign language proficiency, particularly in Middle Eastern and North African languages. Priorities for new equipment include intelligence sensors and platforms (both manned and unmanned), tactical airlift (both fixed wing and rotary wing), specialized precision munitions, and improved communications gear.

Nuclear Forces

The United States retains nuclear forces to deter the use, or threat of use, of nuclear weapons by Russia or China against the United States, its forces, and its allies and partners through threats of limited or large-scale U.S. retaliation. More broadly, by maintaining a U.S. capability for limited use of nuclear weapons at the nonstrategic (theater) and central strategic levels, the United States seeks to induce caution into the actions of decisionmakers in both countries through the possibility of unwanted escalation.

U.S. defense strategy recognizes that both these countries have the ability to overcome the limited air and ballistic missile defenses that protect the U.S. homeland. Successive administrations have therefore chosen to accept a condition of vulnerability to Russian and Chinese retaliatory strikes with strategic nuclear weapons (while not publicly acknowledging our vulnerability to Chinese nuclear attack on our homeland). U.S. deterrent strategy vis-à-vis these states is predicated on the belief that deterrence will hold if the United States can convey that U.S. nuclear forces hold at risk a broad array of highly valued military, political, and economic targets, even if the adversary unleashed a full-scale attack on U.S. nuclear forces first.

U.S. nuclear forces also underwrite extended deterrence relationships with American allies and partners, helping provide assurance against threats posed by regional adversaries. In the case of such states as North Korea and (potentially) Iran, U.S. nuclear weapons undergird American efforts to thwart proliferation. In the event that these states (or others) acquire nuclear weapons, as North Korea already has, U.S. nuclear forces would be called on to deter and, if deterrence fails, to prevent or substantially reduce the effects of their use of nuclear weapons against U.S. forces, allies, or partners. The United States has not been prepared to accept a situation of mutual vulnerability with these nuclear-armed regional adversaries (NARAs).[8]

To operationalize U.S. guarantees of extended deterrence to American allies and partners threatened by NARAs, American forces need to be able to limit damage both by defending against small-scale nuclear-armed missile attacks and by conducting effective counterforce attacks against NARAs' strike capabilities. This calls for warfighting capabilities that combine active ballistic missile defenses and conventional and nuclear strike systems.

U.S. strategic nuclear weapons are deployed on silo-based intercontinental ballistic missiles (ICBMs), nuclear-powered ballistic missile submarines (SSBNs), and long-range B-52 and B-2 bombers.[9] This "triad" of nuclear delivery means has provided the basis for a highly survivable force that, supported by multiple types of surveillance sensors, dedicated command and control and communications assets, and

---

8 For an analysis of the distinctive challenges posed by such states, see David Ochmanek and Lowell H. Schwartz, *The Challenge of Nuclear-Armed Regional Adversaries*, Santa Monica, Calif.: RAND Corporation, MG-671-AF, 2008.

9 In addition, the United States has a force of theater-range, dual-capable fighter-bomber aircraft, which can carry nuclear gravity bombs. U.S. dual-capable F-15Es; F-16s; and, in the future, F-35s, deployed in Europe along with the nuclear gravity bombs they can carry, in combination with allied dual-capable aircraft, play important deterrence and assurance roles for the NATO alliance.

high levels of training and readiness among their operating units, has ensured that no adversary could meaningfully limit damage to his nation by conducting a first strike. Going forward, it may be prudent to retain the triad in some form. It also seems certain that U.S. leaders will want the overall size of the U.S. strategic force to be roughly comparable to that of the Russian Federation, whether the two nations' forces are constrained by mutual agreement or not.

With the exception of the bomber fleet, all major elements of the triad—ICBMs, SSBNs and the missiles they carry, and air-launched nuclear-armed cruise missiles—are nearing the end of their service lives and will have to be retired or replaced. The *Ohio* Replacement Program, which is developing the new SSBN, is by far the most expensive. Estimates vary, but the overall program cost for 12 new SSBNs could exceed $100 billion, with an average cost per ship on the order of $7 billion. High up-front costs notwithstanding, the at-sea portion of the triad has been and will remain the most survivable element of the U.S. deterrent force, and no one in the defense establishment will want to give it up. To date, DoD's long-term plans have not identified funds to pay for production of replacement ships for the *Ohio* class. Replacing or refurbishing the Minuteman ICBM force will be less costly, especially if the silos used to house and launch the missiles are retained.

The primary reason to build a new long-range strike bomber is to improve U.S. conventional power-projection capabilities. These forces need enhanced capabilities to engage and attack a wide range of targets from bases farther removed from the enemy's territory and must be able to overcome sophisticated air defenses, such as those China possesses. The Long-Range Strike Bomber (LRS-B) is being designed to meet these requirements. The marginal cost of also equipping a portion of these new bombers so that they can deliver nuclear weapons will be quite modest. Replacing the current air-launched cruise missile with a new, nuclear-armed standoff missile will also be expensive; however, the replacement missile could be useful in a conventional conflict if outfitted with a conventional warhead. Because bombers and fighter aircraft can deliver nuclear weapons of varying yields and do so flexibly, without having to overfly Russian territory en route to their targets, these aircraft are the most useful elements of the U.S. nuclear forces for addressing threats posed by NARAs.

The Congressional Budget Office (CBO) has examined the currently proposed nuclear force modernization programs and their associated development costs and produced an estimate of total expenditures over the period 2014–2023.[10] These figures include estimates for the *Ohio*-class replacement, the new ICBM, and the new bomber, as well as the new cruise missile and other expenditures for nuclear weapons. Table 3.1 provides a summary of CBO's estimates. Further spending on each of these programs will be required in the years beyond 2023.

Table 3.1
CBO Estimates of Expenditures on Nuclear Systems, 2014–2023 (billions of calendar-year dollars)

Type of Expenditure	SSBNs	ICBMs	Bombers[a]	Other	Total
Department of Defense					
Procurement	33	2	5	—	40
Research, design, test, and evaluation	14	7	12	3	36
Department of Energy					
Weapon activities	6	4	10	—	20
Naval reactors	4			—	4
Total	57	13	27	3	100

SOURCE: CBO, 2013.

NOTE: For SSBNs and ICBMs, the table reflects all the costs expended on the programs during the period. Considerable additional costs will be incurred in the years beyond 2023.

[a] In estimating the costs associated with the bomber leg, all the costs of nuclear weapons or systems, such as the Long Range Standoff Weapon (LRSO), are included, but for the bomber aircraft themselves a special rule is applied: "CBO included in its cost estimates 25 percent of the total anticipated budgets for the B-52 and the LRS-B because that is the fraction of B-52H aircraft that concentrate on the nuclear mission at a given time, in CBO's estimation; in contrast, CBO included 100 percent of the cost of the B-2 and the LRSO" (CBO, 2013, p. 15).

---

[10] CBO, *Projected Costs of U.S. Nuclear Forces, 2014 to 2023*, Washington, D.C., December 2013.

Counterinsurgency and Stability Operations

The current defense plan assumes that the United States is unlikely to conduct protracted counterinsurgency stability operations on the scale of those mounted in Afghanistan, Iraq, and perhaps even the Balkans. As a result, the Army is being cut by nearly 100,000 personnel from its wartime peak of 547,000 active-duty personnel. The U.S. Marine Corps is also losing 20,000 personnel. This decision was premised not on any analysis of possible future needs but rather on the view that such manpower-intensive operations, or at least those of the past decade, have not produced positive results commensurate with their costs. It is certainly true that violent extremist groups continue to operate in both Iraq and Afghanistan, although not in Bosnia or Kosovo. On the other hand, the absence of stabilization efforts can produce even worse results, as evidenced in Syria, Libya, and Iraq following the 2011 U.S. troop withdrawal. As a result, the United States has sent forces back to Iraq, and the Obama administration is under pressure to reconsider its stated intention to withdraw U.S. troops from Afghanistan completely before leaving office. American boots on the ground may or may not be necessary to end the civil war in Syria. Should the Kim regime North Korea collapse, demand for a large-scale stability force may be compelling. Given these and any number of other unforeseeable contingencies, the next U.S. administration may wish to place higher priority on preparing for such missions.

Readiness

A final area of major concern with the currently envisaged defense program is the time it will take to return U.S. forces to a high level of readiness. Readiness, in this sense, refers to the ability of a given force element to execute its full range of assigned tasks. A unit's readiness is a function of the extent to which its personnel are capable and appropriately trained, the maintenance status of its equipment, stocks of expendables (e.g., fuel, munitions), and the availability of spare parts.

As we have seen, the defense strategy calls on U.S. forces to be actively engaged in multiple regions simultaneously. U.S. forces must, among other responsibilities, provide a credible forward presence to deter aggression and assure allies and partners on the Korean peninsula, the Persian Gulf, NATO Europe, and the western Pacific; keep the al Qaeda network and Salafi-jihadist groups under constant pressure through both direct and indirect operations; and be prepared to deploy quickly in response to challenges if deterrence fails. Meeting these commitments requires that a substantial portion of the active-component force (over 80 percent in the case of the Air Force's fighter and bomber squadrons) be trained and ready to deploy in a matter of days.

Overall, the readiness of U.S. forces today is rather poor. The National Defense Panel, which was commissioned by Congress to conduct an independent assessment of the 2014 Quadrennial Defense Review and its findings, assessed that DoD today faces "major readiness shortfalls that will, absent a decisive reversal of course, create the possibility of a hollow force that loses its best people."[11] The reason for this is twofold: More than 13 years of unremitting conflict in Afghanistan and Iraq have placed heavy wear and tear on people and equipment and resulted in large backlogs of equipment that requires depot-level maintenance and repair or replacement. The demands of these conflicts have also meant that many elements of the force have received little or no training for missions other than counterinsurgency and stability operations. This means that, if forces were needed today for combat in, say, Europe or Korea, the President would have to choose between sending troops that were ill-prepared for large-scale maneuver operations and waiting months for them to receive the appropriate training.

These problems were exacerbated by the sudden imposition of sequestration on DoD spending in April 2013. At that point, which was halfway through the fiscal year, the department was compelled to cut $37 billion from its spending for the remainder of the year. The

---

[11] William J. Perry and John P. Abizaid, *Ensuring a Strong U.S. Defense for the Future: The National Defense Panel Review of the 2014 Quadrennial Defense Review*, Washington, D.C.: United States Institute of Peace, July 2014, p. 36.

only practical way to reduce spending that quickly was to cut funds that had been budgeted for training, maintenance, and procurement. The effects were unprecedented and severe: The Air Force, for example, had to ground 13 combat air squadrons for several months and sharply reduced flight training for seven additional squadrons; other services experienced similar disruptions in their maintenance and training activities.[12] By fall 2013, only two of the Army's 43 active-component brigade combat teams were judged to be fully ready and available to execute a major combat operation. Despite subsequent increases in funding for readiness, U.S. forces by and large have still not recovered from the cumulative effects of these stresses.[13] In early 2015, fewer than 50 percent of the Air Force's combat aviation squadrons were rated fully combat capable.[14]

Enhancing Allied Defenses

Countering the threats posed by adversary states is not solely a problem for the United States. In fact, it would be both unwise and infeasible for the United States to attempt to address unilaterally the sorts of challenges outlined above. Allies and partners, particularly those directly or indirectly threatened by adversary activities, have a strong interest in ensuring that their forces can impose a high price on an aggressor and contribute effectively to combined regional operations that may be led by the United States.

With these goals in mind, the proliferation of systems and technologies that are causing U.S. planners such concerns can be turned to our advantage. If allies and partners invest wisely, they can impose smaller-scale A2/AD challenges on the states that are wielding them

---

[12] Mark A. Welsh, "The Impact of Sequestration on National Defense," statement before the Senate Armed Services Committee, November 7, 2013.

[13] Office of the Under Secretary of Defense (Comptroller)/Chief Financial Officer, *United States Department of Defense Fiscal Year 2015 Budget Request Overview*, Washington, D.C.: U.S. Department of Defense, March 2014, p. 3-2.

[14] Jeff Schogol, "Less Than Half of Combat Squadrons Fully Ready for Combat," *Air Force Times*, March 4, 2015.

against us.[15] Taiwan, for example, has both the economic means and the technical and operational savvy to develop, deploy, and operate such systems as short-range unmanned aerial vehicles and antiship cruise missiles, shallow-water mines, rocket artillery, mobile short-range air defenses, and communications jamming gear—all of which, when properly employed, could contribute mightily to an effective defense against invasion.[16] Gulf Cooperation Council countries concerned about aggression from Iran could likewise invest in hardened air bases, mine-sweeping craft, missile defenses, unmanned aerial vehicles, and other capabilities useful in countering conventional and unconventional threats. NATO member states, particularly those in northern and central Europe, should contribute to multinational efforts to strengthen deterrence on the Alliance's eastern flank. Finally, in all regions, regular combined-forces exercises and planning and more interoperable communications networks can help the United States and its allies and partners make the whole of their capabilities as great as the sum of their parts.

Burden-sharing is likely to remerge as an issue, particularly with our European allies. As the figures in Table 3.2 indicate, European defense spending has been in slow decline as a percentage of GDP since the end of the Cold War, while U.S. defense spending has fluctuated, descending rapidly in the 1990s and then rising again the following decade. So far, the crisis in Ukraine and the renewed threat from Russia have had no visible effect on European defense spending, with most governments there fixated on their difficult economic and fiscal situations. American defense spending is currently falling even more rapidly than European spending but is still more than twice as high, as a proportion of GDP, than the European average.

European defense cuts have led to a situation in which it is difficult for European forces to conduct any significant joint combat opera-

---

15 See David C. Gompert and Terrence K. Kelly, "Escalation Clause: How the Pentagon's New Strategy Could Trigger War with China," *Foreign Policy*, August 2, 2013.

16 See Roger Cliff, Phillip C. Saunders, and Scott Warren Harold, eds., *New Opportunities and Challenges for Taiwan's Security*, Santa Monica, Calif.: RAND Corporation, CF-279-OSD, 2011, pp. 7–10.

Table 3.2
Percentage of GDP Spent on Defense

	1990–1994	1995–1999	2000–2004	2005–2009	2010	2013
European NATO members	2.5	2.1	1.9	1.8	1.7	1.6
United States	4.5	3.2	3.3	4.4	5.3	4.4

SOURCE: NATO, Public Diplomacy Division, "Financial and Economic Data Relating to NATO Defence," press release 28, February 24, 2014.

tions without American support. Europe is more vulnerable as a result. The recent NATO summit in Wales recognized the problem and set a goal for defense spending as a percentage of GDP at 2 percent, to be achieved by 2025. Thus far, there are few signs that these budgetary trends will be reversed. As the next administration considers the means needed to support its defense strategy, it will not be able to count heavily on its European allies unless it successfully convinces them to do more in their own defense. This may prove easier if the United States begins to strengthen its force posture in Europe.

Choices for Sizing the Defense Budget

We offer four alternative levels of defense spending and indicate the types of force and capabilities that the United States could sustain at each of these budget levels, described in terms of the ability of each force to address the challenges described in this chapter. Table 3.3 portrays the capabilities that could be sustained under the terms of the BCA (Force I) or the President's FY 2015 budget submission (Force II), and those that could be fielded with substantial and sustained increases in DoD funding above these levels (Forces III and IV). For context, Figure 3.3 shows how U.S. defense spending has risen and fallen over the past 40 years, along with rough projections of spending that would be associated with Forces I–IV.

Table 3.3
Illustrative Consequences of Alternative DoD Funding Scenarios

	Budget Level (FYs 2016–2023)			
	Force I. Budget Control Act (FY 2016 $499 billion)	**Force II. President's Budget (+$115 billion)**	**Force III. "President's Budget Plus" (+$500 billion)**	**Force IV. National Defense Panel (+$900 billion)**
Investment area				
Readiness	Partially ready by 2019	Fully ready by 2019[a]	Fully ready by 2017	Fully ready by 2017
Strategic nuclear forces	?[b]	?[b]	Comprehensive modernization	Comprehensive modernization
CT/SOF	70,000	70,000	70,000	75,000+
Deter or defeat regional adversaries	"2 war minus" posture	"2 war" posture[a]	"2 war" posture	"2 war" posture plus stability operations capacity
Deter or defeat A2/AD	Deteriorating military balance	Selective modernization[a]	Selective modernization	Accelerated modernization
Percentage of GDP in 2024	2.3	2.3	2.5	2.7

[a] Provisions of Force II assume implementation of changes to compensation and health care, base realignment and closure, force structure reductions, and program cuts proposed in the FY 2015 budget submission.

[b] Funds have not been identified to pay for the construction of a new fleet of SSBNs.

Figure 3.3
U.S. Defense Spending as a Percentage of GDP, 1974–2024

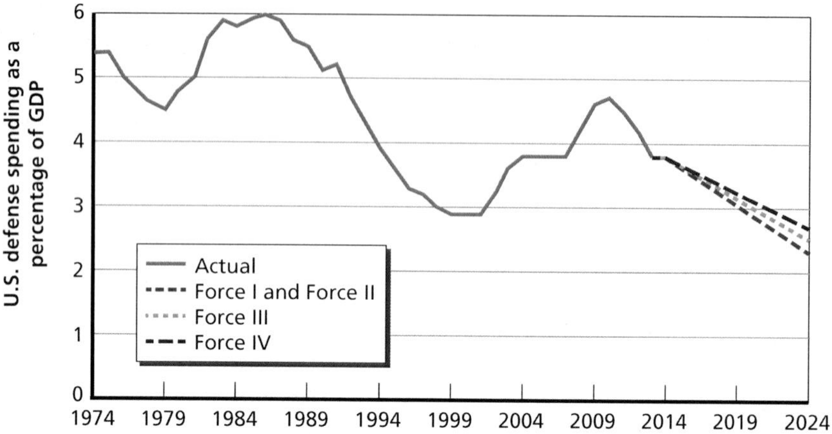

SOURCE: Office of Management and Budget, "Historical Tables," Washington, D.C.: The White House, undated; RAND analysis.
RAND *RR1114-3.3*

Absent unforeseen further demands, funding at the level requested by the current administration (Force II) would allow U.S. forces to reach historically normal levels of readiness by 2019, though somewhat later for the Air Force. This funding level could also sustain the current level of capacity and activities for SOF. However, we note that DoD has not identified funds in the Future Years Defense Program to pay for construction of the new SSBN. This reality, coupled with the likelihood that DoD will not find it possible to reverse decades-long trends in the growth of operations and maintenance costs and the procurement of new platforms, leads us to conclude that, before long, without a significant increase in DoD's top line, decisionmakers will be confronted with painful choices between the aggregate capacity of the general-purpose forces and their modernization. The resulting force would be at risk of falling behind the capabilities of its most modern counterparts or undermining deterrence in one or more regions due to insufficient forward forces and posture.

Not surprisingly, the situation is considerably worse under the BCA caps. As shown in Force I, investments in CT capabilities are sustained, but readiness continues to lag, and the SSBN problem is not solved. In addition to these risks, we project that DoD leaders have to reduce force structure further *and* cut into critical modernization programs, such that U.S. forces would fall behind the capabilities of our most advanced potential adversaries—China and Russia. To put it plainly, America's credibility and influence internationally, the safety and security of its nuclear arsenal, and the viability of its all-volunteer force could all erode if defense spending is held to the levels posited in Forces I or II. In 2024, funding for both Forces I and II is projected to be equal and would constitute approximately 2.3 percent of GDP.[17]

Force III posits an average real increase in DoD funding of approximately $50 billion per year over the period 2015–2023, as compared with the levels permitted by the BCA. Under this scenario, DoD spending by 2024 would constitute approximately 2.5 percent of GDP. At this level of funding, we portray a force that reaches historical readiness levels two years earlier than Force II and, like Forces I and II, sustains present levels of SOF and CT activities. However, Force III also pays for the construction of the initial ships of the modernized SSBN fleet. Force III also invests in the most important initiatives needed for addressing the A2/AD challenge and would provide the wherewithal to substantially strengthen NATO's deterrent posture on its eastern flank. While it will remain essential to garner efficiencies within DoD's infrastructure and personnel accounts, funding DoD at this level would allow more time to enact these politically sensitive reforms.

Finally, Force IV, funded at the level called for by the National Defense Panel, offers all the features of Force III but adds capabilities in three areas: (1) It increases the size of SOF, allowing an expanded level of activity against terrorist groups abroad; (2) it adds capacity, primarily in the land forces, for the sorts of stability operations that U.S. forces have conducted in the Balkans, Iraq, and Afghanistan; and (3) it allows for faster and broader modernization of forces and support facilities called for by the A2/AD threat. At this funding level, the DoD budget would constitute approximately 2.7 percent of GDP

---

[17] Dinah Walker, *Trends in U.S. Military Spending*, Washington, D.C.: Council on Foreign Relations, July 15, 2014, p. 10.

by 2024. Assuming economic growth at the level projected by CBO, all these levels of spending would fall below even the "peace dividend" level of spending of the mid-1990s.[18]

---

[18] CBO, *An Update to the Budget and Economic Outlook: 2014 to 2024*, Washington, D.C., August 2014b, Figure 2.14, p. 46.

CHAPTER FOUR

Counterterrorism

In 2016, a decade and a half will have passed since the tragic attacks of 9/11. For a brief moment in 2011, in the immediate aftermath of Osama bin Laden's death and America's disengagement from Iraq, Americans seemed to put terrorism aside. Then–Secretary of Defense (and former CIA director) Leon Panetta mused publicly that the defeat of al Qaeda was within reach. Terrorism seemed to recede into our memories; Americans felt safer; and Congress and the administration began to consider how to begin to roll back the sprawling CT infrastructure that the nation had spent a decade constructing.

Dismantling most of core al Qaeda in Pakistan was an important milestone in the struggle against global terrorism, but we still face serious threats. We are dealing with a wider dispersion geographically—terrorist safe havens have expanded from Afghanistan before 9/11 to Iraq, Syria, Yemen, Somalia, Libya, and much of the Sahel now. Terrorists have expanded in terms of sheer numbers and recruiting—as evidenced by the massive influx of foreign volunteers to the Islamic State—and in terms of influence as well; terrorists' use of social media and their reach into the West using sophisticated messaging has improved dramatically. Unprecedented instability wracking the Arab world accelerates and incubates militancy. Instead of one hierarchical group with an identifiable leader and a coherent ideology, we now face a multitude of interconnected groups with diverse priorities in a region with declining governance and stability.

The Islamic State's rapid emergence and quick conquest of much of western Iraq and eastern Syria in 2014 surprised the world, its gruesomely publicized and seemingly unbounded brutality continually shocks us. Additionally, its resonance among a sizable multinational gang of followers promises an enduring threat for years to come. ISIS has put terrorism back atop the list of American fears. Rightfully or not, terrorism has returned to the front burner of U.S. foreign policy interests. Polls reveal an American public more concerned about Islamic extremism and more willing to deploy force to deal with it than at any time since 9/11.[1] Now, public fears of terrorism are reaching historic highs. Polling in 2011 showed one-third of Americans worried about becoming a victim of terrorism (down by one-half from two years before). In early 2015, nearly three-quarters of Americans thought a catastrophic attack against the United States would take place unless we take action to stop the Islamic State.

The current administration has sought to move from counterinsurgency to more narrow, precision-targeted counterterror strikes as its preferred method for dealing with foreign terrorist threats. The results have been mixed. Many terrorists have been killed and many attacks disrupted, but the overall threat has continued to expand. The next administration will consequently face several interrelated issues and difficult choices:

- What is the right balance between tactical and strategic CT efforts?
- What will be needed to defeat the Islamic State, first in Iraq and then, more daunting, in Syria?
- How do we approach picking partners and prioritizing enemies in what is an increasingly complicated CT landscape?
- Should we rebalance where we need to be on the spectrum of privacy and security?

---

[1] Pew Research Center, "Growing Support for Campaign Against ISIS—and Possible Use of U.S. Ground Troops," February 24, 2015. Twice as many Americans approve as disapprove of the U.S. military campaign against ISIS. Even the idea of sending ground troops to Iraq draws significantly more support than it did several months ago. Fear of Islamic extremism—both overseas and domestically—has jumped from a low point in 2011 to its highest level in a decade.

Tactical Versus Strategic Counterterrorism

The first choice for the upcoming administration may be whether and how to shift the balance between the day-to-day CT campaign and more-strategic efforts to counter violent extremism. Tactical efforts have urgency and receive the bulk of resources and priority—these operations include lethal strikes against terrorist leaders and massive law enforcement and intelligence efforts to uncover, understand, and disrupt ongoing plots. Critics of the tactical emphasis have argued that, without a longer-term approach designed to stem the appeal of terrorism, we will forever be stuck in a cycle of killing one generation of terrorists while creating the next.

Tactical Emphasis

This is less a choice than a continuation of current priorities—focus on preventing attacks and eliminating terrorist leaders. We have done this with remarkable success since 9/11. Most of the original plotters of the 9/11 attacks are dead. We have stopped almost every plot against the homeland. How? Through enormous investment in strong defenses and aggressive offenses and a minute focus on the enemy. The administration, Congress, and the American people have set the threshold for accepting attacks at near zero and expect near perfection from the CT community. The few successful attacks—the Boston Marathon, Fort Hood, and the near-success of the underwear bomber—all resulted in a frenzy of finger pointing and criticism. These defensive measures have proved hugely expensive while largely containing but not fundamentally reducing the threat.

Strategic Shift

This choice entails gradually reducing our focus on the tactical (in particular, the lethal aspects that have eliminated scores of terrorist leaders but do not seem to have eliminated the threat) toward more strategic CT programs, including more and more aggressive community outreach, counter messaging, counter radicalization, and other measures to combat violent extremism. Funding and priority have never matched the rhetorical support for providing positive alternatives to those most at risk of radicalization and recruitment, countering violent extremist narratives and messaging, and working with partners both at home and abroad to address the drivers of radicalization.

What Will Be Needed to Defeat ISIS?

The war against ISIS is likely to see more success in Iraq than in Syria. By the end of 2016, it is conceivable that some combination of U.S. airpower with Kurdish and Shi'a forces, retrained Iraqi Army units, and turned Sunni tribesmen fighting on the ground will have recaptured much of the Iraqi territory that ISIS took in summer 2014, including Mosul and other large population centers. Even in that best-case outcome, though, it seems highly unlikely that all of Anbar province will be returned to central government control and even less likely that Syrian territory will be wrested from the grip of ISIS.

In Iraq, the harder challenges involve politics and governance more than CT and will require outreach and inclusivity not yet shown by the Shi'a government in Baghdad. To keep the Sunnis from siding with ISIS, Baghdad must give them a better alternative. That has been too slow in coming and will require persistent and aggressive U.S. engagement with the government in Baghdad.

In Syria, the challenges are more difficult and daunting and the choices less attractive:

- Is there a moderate Syrian opposition that can succeed in battle against ISIS? The Kurds have shown it can be done, but they have decades of experience, an organized military structure, and unity of purpose. The Syrian opposition has none of these, is still mired in debate, and has little military capacity to bring to bear. Is the training program for the Syrian moderate opposition realistic?
- More coalition airstrikes against ISIS will benefit its enemies in Syria, who happen to be U.S. enemies as well. Both the Assad regime and al Qaeda–affiliate Jabhat al-Nusrah, probably the most powerful force on the ground after the regime and ISIS, will emerge stronger as ISIS becomes weaker. Are U.S. policymak-

ers willing to indirectly bolster the Assad regime in their efforts to degrade ISIS? If ISIS is degraded in Syria, are they willing to engage Jabhat al-Nusrah and other militants militarily to stem their growth and prevent them from gaining strength?

All Air

Resisting calls from the region and from some in the United States to ramp up support to anti-ISIS forces (the Kurds, Shi'a, so-called Syrian moderates, even the Syrian regime), this choice would limit U.S. military involvement to an aggressive air campaign in support of other non-U.S. forces on the ground. Training and support are elements of this approach that also reduce the likelihood of heavy American casualties or finding ourselves committed to a long-term ground campaign.

All In

Critics of the air-only approach, both domestic and among regional allies, such as the Kurds, and critical Arab partners, such as the Saudis and Jordanians, have ramped up their calls for more U.S. involvement. They question the policy of no boots on the ground. They argue that air attacks can keep ISIS contained but will never defeat it. Few argue for brigades or divisions of U.S. soldiers, but many argue that more logisticians, more trainers, better intelligence support, and spotters for the air campaign are necessary to make it more effective.

Picking Partners and Identifying Enemies

This is a choice between relying on carefully selected local partners, on the one hand, or putting the United States front and center in the fight against extremism, confronting the threat wherever it emerges, and pulling no punches in labeling the threat "Islamic" extremism. Most of our Muslim allies, along with our own Muslim population, have lobbied against singling out Islam as the problem. They argue that the United States is an essential partner and can provide important support in the struggle but that the phenomenon we are witnessing in the Muslim world is a struggle within Islam, not between Islam and the West. However, former Defense Intelligence Agency director General Michael T. Flynn and other prominent observers have suggested that the fight against ISIS and al Qaeda, indeed against Islamist radicals broadly, is a generational struggle against a global enemy that the United States must lead. This camp argues that, as long as the United States stands on the sidelines, reluctant to clearly identify and engage the enemy head on, the threat will increase.

Engage Reluctantly, Partner Carefully, and Target Only the Most Dangerous Threats

This choice rejects the recommendation that America must confront Islamic extremism wherever it surfaces and lead the fight wherever it occurs. Instead, the administration would apply selective criteria to countries with which it is willing to partner—basing decisions on potential partners' commitment to democracy, human rights, and the appropriate conduct of their militaries. At the same time, it would prioritize enemies by the imminence (and scale) of the threat they pose to the United States (first to the homeland, then to its interests overseas).

Broaden Our Partnerships and Engage a More-Diverse Enemy

This approach would entail embracing countries threatened by Islamic extremists, with less regard to their humanitarian records and whether the extremists threaten the U.S. homeland directly or not. As Flynn has said,

> We are in a global war with a radical and violent form of the Islamic religion . . . we must engage the violent Islamists wherever they are, drive them from their safe havens, and kill them. There can be no quarter and no accommodation.[2]

This choice will force some tough decisions: How might the United States consider a closer partnership with Nigeria in that country's battle with the ruthless and bloody Boko Haram? In Algeria, the shadowy army-led power behind the regime (known as *le pouvoir*, or "the

[2] Kimberly Dozier, "Former DIA Chief Flynn Calls for Global War on Islamic Extremists," *Daily Beast*, February 13, 2015.

power") wields almost complete control and has little regard for collateral damage but has been ruthlessly effective against its militant enemies. How might the United States help Algeria in its campaign against al Qaeda in the Islamic Maghreb without contributing to the environment that encourages militancy?

Rebalancing Civil Liberties and Security

In the immediate aftermath of 9/11, law enforcement and intelligence agencies were granted broad new authorities to access, collect, and exploit electronic data to help unravel the al Qaeda network, penetrate and prevent plots, and identify possible terrorist operatives. Bulk collection became the preferred approach, vacuuming up vast amounts of data and sorting through it for relevant connections. Protections for data on U.S. persons were incorporated, but top priority was given to uncovering potentially threatening information.

More than a decade after the 9/11 attacks and against the backdrop of the perceived successes in CT and a sense of greater security, the revelations of Edward Snowden produced a strong public backlash and led to demands to rescind the authorities that were granted in 2001. Reforming these authorities became the priority of many in Congress and of the administration itself. In January 2014, President Obama presented his response to the recommendations from his Review Group on Intelligence and Communications Technologies. The President's approach was cautious, balancing the need for surveillance, but revealed a change in priorities, with a clearer emphasis on privacy, putting in place more oversight and restrictions.[3]

Favor Civil Liberty

Under this alternative, the next administration would strengthen initial steps taken by the Obama administration and further rescind authorities granted in the Patriot Act and subsequent revisions. Polling shows

---

[3] Barack Obama, "Remarks by the President on Review of Signals Intelligence," Washington, D.C.: The White House, Office of the Press Secretary, January 17, 2014a.

a vocal and sizable minority (and a majority of young voters) remains highly skeptical of government surveillance. This constituency argues that, despite what the government might say about the threat, there is a clear need to strengthen protections to safeguard civil liberties; reduce the National Security Agency's (NSA's) ability to operate in secrecy; and limit the government's ability to collect, monitor, and store communications, both content and metadata, particularly on U.S. citizens.

Favor Security

In this option, the next administration would reverse some of the new restrictions on intelligence gathering and storage, returning to a pre-Snowden, post-9/11 posture of putting security first. The rise of ISIS and a heightened sense of threat could strengthen those who call for more surveillance authority. In the aftermath of the next attack in the United States, momentum will inevitably shift to a more-aggressive intelligence approach. On the eve of the underwear bomber's attempt to down an incoming aircraft, criticism focused on too much surveillance, too many watch list entries, and too much budget allocated to CT. The day after the attempt, Congress was highly critical of the Intelligence Community's performance, the result being a campaign that resulted in more money for data acquisition, a dramatically expanded watch list, and a more-flexible remit for the Intelligence Community in terms of acquiring, storing, and analyzing information.

Whichever of the above approaches is chosen, there will need to be an effort to harmonize collection and privacy-protection efforts with at least close allies. In the absence of some commonality of approach, Western intelligence services and legal regimes could find themselves working at cross purposes, hindering their effectiveness and leaving dangerous gaps in their threat coverage.

CHAPTER FIVE

Cybersecurity

In the 25-year history of the Internet, no cyberattack is known to have killed a human being.[1] No country has declared cyberwar or even admitted to sponsoring a cyberattack. Nevertheless, the damage caused by human malice enhanced by digital capabilities is mounting. This includes the *physical* disabling or destruction of computers and networks and the *economic* losses due to cybercrime and theft of intellectual property—estimated to be at least $345 billion per year worldwide.[2] These losses have, in turn, prompted private companies, governments, and individuals to invest more in preventing future attacks; global spending on cybersecurity is now estimated at about $70 billion per year.[3] Damages also include *political* fallout from the theft of U.S. government secrets and exposure of surveillance and espionage activities. Also, there are *human* costs, in reputations and livelihoods destroyed by invasions of online privacy, as well as identity, tax, and medical fraud.

Although Internet freedom, cybersecurity, and cyberwar policies may not loom large in the 2016 U.S. presidential campaign, they will present significant challenges to the next administration. More than three billion Internet users—two billion more than a decade ago—

1 Martin Libicki, *Cyberdeterrence and Cyberwar*, Santa Monica, Calif.: RAND Corporation, MG-877-AF, 2009, p. 112.

2 Center for Strategic and International Studies, *Net Losses: Estimating the Global Cost of Cybercrime*, Washington, D.C., June 2014.

3 Martin Giles, "Defending the Digital Frontier," *The Economist*, July 12, 2014.

now have access to increasingly valuable content and to each other, in real time, from almost anywhere on earth.[4] Yet with greater connectivity comes greater vulnerability. The next administration will need to prioritize defending U.S. government and military networks and cyber operations, as well as critical infrastructure, against a range of attackers that may at any time include sophisticated rival states, terrorists, politically motivated hackers, financially motivated criminals, and high-tech thrill seekers. At the same, it will have to defend the national interest in preserving an open, interoperative, secure, reliable, and global network.

Domestically, policymakers will need to step up implementation of policies aimed to strengthen U.S. government defense of the private sector against cyber espionage, exploitation, and attack, including critical infrastructure that is privately owned.

Internationally, if the United States aspires to project global leadership on these issues, it will need to overcome the skepticism engendered by leaks about its surveillance practices; articulate credible and thoughtful policies for cybersecurity, Internet governance, online surveillance, and deterring cyberwar; and win greater domestic and international support for these policies.

U.S. Strengths and Vulnerabilities

The United States, as the global leader in technology, has the most to lose of any country from theft of its intellectual property. As the clear leader in ICT and a major ICT exporter, it also benefits from its intellectual and commercial dominance in this area. But its ever-growing dependence on advanced technologies—Internet-based and otherwise—increases its vulnerability to technological disruption. The 2015 DoD Cyber Strategy includes the startling statement that, since 2013, the Director of National Intelligence has ranked the threat of cyberattack as "the number one strategic threat to the United States, placing it

---

4 DoD, *The DoD Cyber Strategy*, Washington, D.C., April 2015, p. 1.

ahead of terrorism for the first time since the attacks of September 11, 2001."[5]

In an April 2015 speech at Stanford University, Secretary of Defense Ash Carter announced a new drive for public-private partnerships to keep the United States ahead of its adversaries in developing the technologies needed to thwart cyber-enabled espionage, sabotage, extortion, and attack by hostile states, criminals, or terrorists.[6]

Whatever the Obama administration's progress toward these ends, the next administration and Congress will continue to face difficult choices about how to protect online liberty, privacy, and security, all of which have been compromised. Contrary to conventional wisdom, privacy and security online are not always in conflict; better cybersecurity practices also protect privacy. However, the tension between protecting against attacks by terrorists, cybercriminals, or nation-states and preserving the privacy of ordinary Americans will not be easily resolved.

For example, the next administration may have to balance demands to preserve anonymity and free movement on the Internet with efforts to reduce the "attribution problem," in which cyberattackers remain difficult or even impossible to identify. It will continue to face demands for more transparency about offensive cyber and domestic and international surveillance operations and charges of hypocrisy if it rejects them. At the same time, since cyberattackers can surprise their targets by exploiting unknown or unpatched vulnerabilities, the U.S. government and the private sector will often need to act in secret to conceal and repair vulnerabilities, identify or punish attackers, or track increasingly tech-savvy terrorists. This requires a level of trust between government and tech companies, a relationship that many view as damaged. These dilemmas are by no means exhaustive, but they illustrate the political complexities that compound the technical difficulties of cybersecurity.

---

5 DoD, 2015, p. 9.

6 Ash Carter, "Remarks by Secretary Carter at the Drell Lecture Cemex Auditorium, Stanford Graduate School of Business," transcript, Stanford, Calif., April 23, 2015.

Things to Come

In nearly every realm of computer- and network-based activity, cycle times are compressed, and everything seems to happen faster—sometimes many times faster—than humans can effectively respond. We can currently discern only some of the ways in which technology is accelerating. One known factor is the proliferation of autonomous systems. Led by the military needs of the past several decades, remotely piloted systems, such as unmanned aircraft, have taken over many dangerous, dirty, or dull tasks of soldiers, sailors, and aviators. While military needs will continue to drive niche applications, and as autonomous capabilities begin to get integrated, both remote and autonomous capabilities are also being commercialized for a wide range of civilian applications to reduce the number of routine tasks that need to be done by humans.[7]

Another possible accelerant of technological and societal change will be the advent of the so-called Internet of Things (IoT). In the coming years, our physical infrastructure—including billions of gadgets, objects, and all manner of machines—will be outfitted with smart sensors and hooked up to the Web, where they will begin uploading data about themselves and their surroundings and downloading data to enhance their own functions.[8] At least another billion people are expected to come online, but so are as many as 50 billion "things."[9] By 2020, online devices are projected to outnumber human users by a ratio of six to one.

---

[7] The commercial use of drones faces negative public perceptions because this technology developed for military and intelligence purposes has raised privacy and safety concerns within the United States. Amazon, for example, is testing delivery drones in Canada after failing to obtain permission from the United States. See Ruth Reader, "Amazon's Drones Could Track You Down to Deliver Your Goods," *VB News*, May 8, 2015.

[8] Keith Mercier, "The Internet of Things Will Transform Retail as We Know It," *Forbes*, January 12, 2015.

[9] Most of the "things" on the Internet will be controlled by the private sector, raising interesting new questions about the scope of the U.S. government's responsibility to protect them. See Department of Homeland Security, "What Is Critical Infrastructure?" web page, October 24, 2013. For more about the IoT, see Cisco, "Seize New Product and Revenue Opportunities with the Internet of Things," web page, undated, and Mercier, 2015.

This IoT could give us unprecedented situational awareness, efficiencies, and consumer pleasures but could also introduce new risks to privacy and security by increasing the opportunities for a cyberattacker to use virtual capabilities to cause physical damage to U.S. infrastructure or injury to American citizens.

Since the future will bring a vastly expanded cyber domain, the cyber attack surface will correspondingly increase, providing military attackers, as well as those bent on espionage, extortion, or sabotage, more opportunities to exploit both engineering and human weaknesses. One scenario with traction among security authorities is an attack against the nation's critical infrastructure that would have a debilitating effect on military or economic security, national public health or safety, or any combination thereof.[10]

The United States can and will make the case that all nations are made vulnerable by the threat of cyberattack or cyberwar. However, at the moment, the most technologically advanced and wired nations—the United States and its friends and allies—have the most to lose from cyberattacks. Therefore, addressing the current state of weakness in cyber deterrence will be critical.

Addressing Vulnerabilities

In the future, as in the past, most attackers will exploit poor online hygiene.[11] Failure to employ well-known—and comparatively simple and inexpensive—security best practices is the key cyber vulnerability

[10] Department of Homeland Security, 2013. In response to such scenarios, the National Institute for Standards and Technology developed the *Framework for Improving Critical Infrastructure Cybersecurity* (Version 1.0, February 12, 2014).

[11] Practices that can be considered good online hygiene include, among others, avoiding malicious email attachments, compromised websites, or infected media (e.g., thumb drives); employing antivirus and antispyware scanners; updating applications, software, and operating systems within 48 hours of patches becoming available; securely configuring systems and devices; securing browsers and browser add-ons; encrypting and backing up data; securing wireless networks; protecting and limiting administrative accounts; employing effective firewalls; monitoring software audits; and conducting continuous (automated) vulnerability assessment and remediation.

and the reason for a vast majority of cyberattacks.[12] The DoD Cyber Strategy warned the private sector that

> [w]hile the U.S. government must prepare to defend the country against the most dangerous attacks, the majority of intrusions can be stopped through relatively basic cybersecurity investments that companies can and must make themselves.[13]

Our national, community, commercial, and personal lives will not be secure online without a fundamental change to our culture of cyber hygiene, just as cultural changes were needed to get people to practice germ hygiene and wear safety belts. This will require changes in education, the attitudes of business leaders toward proactive investment in hardening their systems and their products, widespread training of students in computer security practices, and careful engineering of the "things" to be connected to the network. At the same time, some critical computers and controls will have to be unhooked from the network entirely, a practice known as air gapping.

Absent effective national leadership, policy formulation, and a legal framework, the ability to protect power, water, transportation, communication, and financial services, as well as government operations, will be uncertain in times of calm and precarious in times of conflict. The current administration has laid a foundation and started down a path of greater cyber resiliency and crisis responsiveness, but progress has been limited.[14]

The next administration will need to improve its capabilities by hardening its systems and networks against the likelihood of attack, investing in backup measures to reduce the impact of an attack, and

---

[12] Conventional wisdom places this figure at ~80 percent, but a multiyear cyber forensic study led by Verizon indicated the number to be 97 percent. See Verizon, *2012 Data Breach Investigations Report*, March 2012. Recent high-profile web breaches indicate the vast majority of attackers exploit poor cyber hygiene.

[13] DoD, 2015, p. 5.

[14] U.S. Government Accountability Office, *Federal Information Security: Mixed Progress in Implementing Program; Improved Metrics Needed to Measure Effectiveness*, September 26, 2013.

budgeting to build back better capabilities than those that may be destroyed. Depending on its calculations of the likelihood of severe cyberattacks, the next administration may choose to spend more on cybersecurity and resiliency now, or risk spending much more later to deal with the consequences of a successful cyberattack.

Privacy Versus Security

The American public is increasingly concerned about civil liberties, privacy, and security in the digital world.[15] Many have had their financial or personal data compromised or stolen. Others worry that their private online activities may be captured, stored, and monitored by the government without their knowledge and without a warrant. Some lawmakers promise to block reauthorizing the Patriot Act or curtail the authorities it granted.

Abroad, the United States has drawn sharp criticism after revelations that it has been collecting vast amounts of data on foreign citizens by intercepting Internet communications and the conversations of foreign leaders—although many other countries do the same.[16] The United States is viewed as hypocritical in its support for Internet freedom, where the U.S. champions an open, neutral, and nongovernment controlled Internet for all users—even those within repressive regimes. American ICT companies have complained that the loss of trust has undercut their international markets and exports, as wary consumers and protectionist governments find rationale to avoid U.S. digital products and services. In response, ICT companies are developing stronger and more-pervasive encryption that is out of the reach of law enforcement and intelligence services and that cannot be unlocked by the company itself. As a result, companies will be unable to comply with

---

[15] President Obama noted public attitudes toward perceived loss of control over personal data in a January 2015 speech at the Federal Trade Commission. See Barack Obama, "Remarks by the President at the Federal Trade Commission," Washington, D.C.: The White House, Office of the Press Secretary, January 12, 2015.

[16] Angelique Chrisafis, "France 'Runs Vast Electronic Spying Operation Using NSA-Style Methods,'" *The Guardian*, July 4, 2013.

national security letters that the U.S. government used in the past to demand user data from the companies.[17] This will create a much more challenging intelligence collection environment.

Prioritizing Privacy

The next administration may want to reverse the erosion of trust in government surveillance practices. Working with Congress, industry, legal, and advocacy leaders, it might achieve some degree of consensus on the collection, retention, use, or dissemination of personal information and communications of U.S. citizens. With so-called minimization guidelines, it might limit the information collected to that directly related to specific threats and might agree to review the processes, practices, and execution of operations. The administration could also ask Congress to amend the Foreign Intelligence Surveillance Act (FISA) to increase transparency and civil liberties protections in the FISA courts.

Prioritizing Security

Alternatively, the next administration might make the case that Americans should trade some of their privacy for greater protection against terrorism and other security threats. It could point to the example of France, whose parliament recently voted for a sweeping expansion of government surveillance authorities following the January 2015 attack by French al Qaeda operatives on *Charlie Hebdo*, a satirical magazine in Paris.[18] At the very least, the administration would seek support for continuing current surveillance and data retention rules.[19] It might seek congressional authorization for more-robust collection efforts and the requirement that companies retain user data for longer periods to aid in investigations.

---

[17] For more about the tensions over encryptions, see Ellen Nakashima and Barton Gellman, "As Encryption Spreads, U.S. Grapples with Clash Between Privacy, Security," *Washington Post*, April 10, 2015, and Joel Hruska, "The NSA Wants 'Front Door' Access to Your Encrypted Data," *ExtremeTech*, April 13, 2015.

[18] Alissa J. Rubin and David E. Sanger," Familiar Swing to Security Over Privacy After Attacks in France," *New York Times*, May 6, 2015.

[19] Executive Order 12333, United States Intelligence Activities, December 4, 1981.

Limiting the Likelihood of Cyber War

While rare, destructive state-originated cyberattacks (as opposed to cyber espionage) have become an instrument of national power. Examples include Stuxnet, the alleged U.S.-Israeli attack on the Iranian nuclear program; Chinese exploitation of networks around the world;[20] Russians taking down networks in Estonia,[21] Georgia,[22] and Ukraine;[23] North Korean attacks on commercial systems in South Korea[24] and the United States,[25] and Iranian attacks on U.S. financial institutions and on the Saudi energy sector.[26] These offensive attacks went beyond intelligence collection, malware installation, or data exfiltration. They suggest that cyber weapons may be chosen for future confrontations and conflicts.

U.S. cyberwar capabilities rival or exceed those of today's potential adversaries.[27] Among its strengths are a superior ICT industrial base, best-in-breed surveillance and collection agencies, and the emerging ability to conduct integrated surveillance-attack operations in U.S. Cyber Command, which is colocated with the NSA. However, there

---

[20] For more on Chinese cyber activities, see Mandiant Intelligence Center, *APT 1: Exposing One of China's Espionage Units*, undated; CrowdStrike, *Putter Panda: PLA 3rd Department 12th Bureau Unit 61486*, June 9, 2014; and Mark A. Stokes, Jenny Lin, and L. C. Russell Hsiao, *The Chinese People's Liberation Army Signals Intelligence and Cyber Reconnaissance Infrastructure*, Project 2049 Institute, November 11, 2011.

[21] Jason Richards, "Denial-of-Service: The Estonian Cyberwar and Its Implications for U.S. National Security," *International Affairs Review*, Vol. 18, No. 2, 2009.

[22] John Markoff, "Before the Gunfire, Cyberattacks," *New York Times*, August 12, 2008.

[23] Ellen Nakashima, "Russian Hackers Use 'Zero-Day' to Hack NATO, Ukraine in Cyber-Spy Campaign," *Washington Post*, October 13, 2014.

[24] Tania Branigan, "South Korea on Alert for Cyber-Attacks After Major Network Goes Down," *The Guardian*, March 20, 2013.

[25] David E. Sanger and Martin Fackler, "N.S.A. Breached North Korean Networks Before Sony Attack, Officials Say," *New York Times*, January 18, 2015.

[26] Kim Zetter, "The NSA Acknowledges What We All Feared: Iran Learns From U.S. Cyberattacks," *Wired*, February 10, 2015.

[27] Dan Goodin, "How 'Omnipotent' Hackers Tied to NSA Hid for 14 Years—and Were Found at Last," *Ars Technica*, February 16, 2015.

is disagreement over whether the United States should use its offensive capabilities against adversaries or be circumspect to avoid escalation into cyberwar.[28]

Given the nature of competition in military and civil technology, U.S. superiority in computer network exploitation and attack may erode or even disappear over the coming decades.[29] But U.S. conventional military superiority will likely endure, providing ample motive for adversaries to take to cyberspace, not the physical world, to wage asymmetrical attacks. To leave burgeoning digital technologies unconstrained by the kinds of arms control agreements that have been put on the use of nuclear and conventional weapons is to increase the likelihood that cyberspace may become a future battlefield of choice.

Prioritizing the Development of Norms

A rules-based international agreement that distinguishes acceptable from unacceptable online behaviors could favor the United States and its allies—even if some states sometimes choose to flout the norms. For precisely that reason, it will remain difficult to persuade U.S. adversaries to agree to such norms.

International norms circumscribing the behavior of states are notoriously difficult to establish and enforce. At present, most analysts see little hope of expanding cyber treaties and legally binding mechanisms, given this opposition, problems of attribution, the large spectrum of cyber parties, and verification difficulties. Nevertheless, the process of developing, advocating, and establishing international

---

[28] The latter view has been promoted by Edward Snowden, the former NSA contractor viewed by some as a whistleblower and by others as a traitor: James Bamford and Tim De Chant, "Exclusive: Edward Snowden on Cyber Warfare," interview, NOVA, PBS, January 8, 2015.

[29] The DoD Cyber Strategy states

> Russia and China have developed advanced cyber capabilities and strategies. Russian actors are stealthy in their cyber tradecraft and their intentions are sometimes difficult to discern. China steals intellectual property (IP) from global businesses to benefit Chinese companies and undercut U.S. competitiveness. While Iran and North Korea have less developed cyber capabilities, they have displayed an overt level of hostile intent towards the United States and U.S. interests in cyberspace. (DoD, 2015, p. 9)

norms for cyberspace might, in the long run, provide the next administration a viable means to lessen risks and isolate violators.

At the minimum, this would involve defending the status quo. The United States could continue to maintain existing systems of Internet governance, including the nonprofit Internet Corporation for Assigned Names and Numbers (ICANN), despite the objections of Russia, China, and other countries to what they view as inappropriate U.S. control over the global Internet. Washington may continue to attempt to establish and shape new international norms and the Internet policies of other states using multistakeholder initiatives, such as the Internet Governance Forum, and reject attempts to put the Internet under the control of any United Nations body.

To date, there has been some initial, but very limited, progress in developing international norms with the *Tallinn Manual on the International Law Applicable to Cyber Warfare* (though Beijing has not agreed to this), the Budapest Convention on Cybercrime (though both China and Russia reject it), and the U.S. commitment to international cooperation in its International Strategy for Cyberspace.[30] The next administration might embark in a reinvigorated, and continuous, process of engagement and dialog regarding international norms.

The benefits could include

- creation of patterns of engagement and expectations to provide opportunities to resolve crises and disagreements as they arise
- increased understanding of mutual risk
- development of a "socializing" effect and domestic pressure to encourage others adhering to the norms
- providing a diplomatic mechanism for raising awareness of norm violation[31]

---

[30] NATO Cooperative Cyber Defence Centre of Excellence, *Tallinn Manual on the International Law Applicable to Cyber Warfare*, Cambridge, UK: Cambridge University Press, 2013; Council of Europe, Convention on Cybercrime, Budapest, November 23, 2001.

[31] Royal United Services Institute, *Cyber Norms of Behaviour: Executive Summary*, London, undated.

- eventually, developing a means for effectively assigning responsibility for policing online actions that violate international law or norms; this would include techniques to spotlight violations without compromising perishable surveillance capabilities.

To the extent that compliance with newly established norms can be verified and assured, these would reduce risk and enhance stability in times of crisis. Norms that limit U.S. options but not those of potential adversaries could, however, have the opposite effect. Thus, restraints agreed among parties to any new norms should not protect those who do not agree—or do not abide by their agreement—from retaliation.

Unrestrained Competition

The next administration might chose to rely on sustained American technological superiority over all potential foes to defend against and deter attacks upon U.S. networks. It might want to send a clear, unambiguous, and credible signal that U.S. cyber capabilities and operational art would prevail in any conflict, that Cyber Command could identify and punish any cyberattacker, and that it has the capability to deny any benefit the attacker might seek.

This could involve ensuring the constructive maturation of Cyber Command and its relationship with the NSA. The administration could leverage the command's integration of cyber offense, defense, and intelligence—an advantage that the United States enjoys over most other military cyber powers. If cyberattacks or damage escalate, the next administration might also choose to prioritize cyber superiority over political considerations.

CHAPTER SIX

Climate Change

The climate change challenge is a byproduct of the world's economic success. In the 70 years since the end of the Second World War, the global economy has grown more than tenfold.[1] In the past two decades alone, one billion people have risen out of extreme poverty.[2] Fossil fuels—oil, gas, and coal—have provided most of the energy needed to propel this growth,[3] but burning these fuels emits gases, primarily carbon dioxide, that in the atmosphere help regulate the Earth's temperature. Today, atmospheric concentrations of greenhouse gases are 40 percent higher than at the start of the industrial revolution—400 parts per million today, compared with 280 parts per million in the 1780s, with three-quarters of that rise occurring over the past 45 years (see Figure 6.1). Global mean surface temperature is about 1°C warmer.[4] By mid-21st century, if the Earth's roughly 9 billion residents achieve desired standards of living while maintaining the historic reliance on fossil fuels, the atmospheric concentrations of greenhouse gases could

---

[1] Angus Maddison, *The World Economy: A Millennial Perspective*, Paris: Organisation for Economic Co-operation and Development, 2001.

[2] "Towards the End of Poverty," *The Economist*, June 1, 2013.

[3] Vaclav Smil, *Energy Transitions: History, Requirements, Prospects*, Santa Barbara, Calif.: Praeger, 2010.

[4] Intergovernmental Panel on Climate Change, *Climate Change 2013: The Physical Science Basis*, Cambridge, UK: Cambridge University Press, 2013.

Figure 6.1
Global Atmospheric Greenhouse Gas Concentration

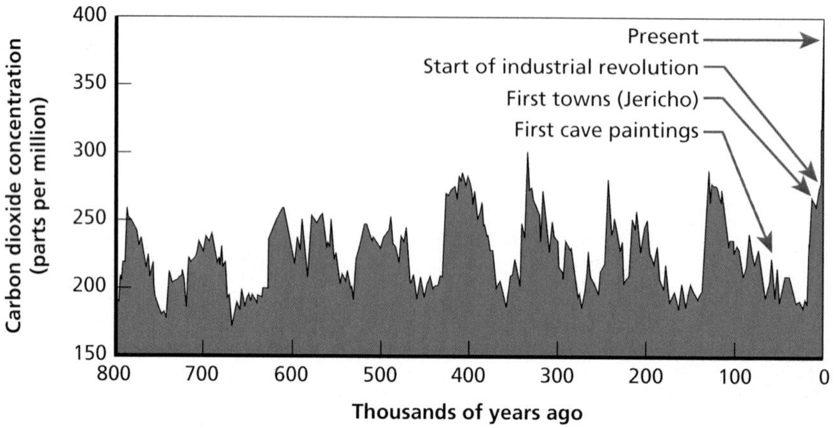

SOURCE: Andrew Freedman, "The Last Time CO2 Was This High, Humans Didn't Exist," *Climate Central*, May 3, 2013.
NOTE: Greenhouse gas concentrations are now higher than they have been in nearly a million years and are rising faster than they have in the entirety of human civilization.
RAND *RR1114-6.1*

roughly double from preindustrial levels,[5] with potentially significant deleterious effects.

The United States faces two broad choices in addressing climate change both at home and abroad: how much to mitigate—that is, reduce—greenhouse gas emissions and how much to adapt to the potential impacts of a changing climate. No matter how much the United States and other nations reduce emissions now, a significant amount of climate change is already "baked into" the system, for two undisputed scientific reasons: The oceans take many decades to warm, and many greenhouse gases persist in the atmosphere for centuries. Even if greenhouse gas emissions were to cease tomorrow, the climate would continue to change for several decades to come.

While there is overwhelming scientific evidence that human activities are increasing the levels of greenhouse gases in the atmo-

[5] Intergovernmental Panel on Climate Change, *Climate Change 2014: Mitigation of Climate Change*, Cambridge, UK: Cambridge University Press, 2014b.

sphere and that this increase is changing the earth's climate, the precise future impacts remain shrouded in uncertainty.[6] Scientists can predict some trends with confidence: It will get hotter, and the seas will rise. Other trends—such as the intensity of storms and how rain pattern will shift—are much harder to forecast. But the impacts of climate change will also depend on whether humans respond well or poorly to its disruptions. The timing and extent of human adaptation is at least as important as the magnitude of climate change itself.

For political, economic, strategic, and humanitarian reasons, the next administration may decide that the United States has a national interest in advancing some combination of mitigation and adaptation strategies at home and around the world. Limiting climate change to acceptable levels may require reducing net global emissions to near zero. Breaking the historic connection between greenhouse gas emissions and economic growth would entail radical changes in transportation, energy, and water systems; agriculture; infrastructure; buildings; and behaviors across all sectors of the global economy.

Decisions made today anywhere in the world about infrastructure investments will have long-lasting implications for both future emissions and future resiliency, defined as the capacity of places to bounce back quickly from disasters or build better infrastructure and systems than what was destroyed.[7] But because we do not know how successful global efforts to reduce emissions may be, it is difficult to predict whether adaptations in advanced societies will be sufficient to avert more drastic consequences. The developing world is another story. In the short term, the worst consequences are expected in the least developed countries, which are both more vulnerable to climate change and less able to adapt.

---

[6] C. P. Weaver, R. J. Lempert, C. Brown, J. A. Hall, D. Revell, and D. Sarewitz, "Improving the Contribution of Climate Model Information to Decision Making: The Value and Demands of Robust Decision Frameworks," *Wiley Interdisciplinary Reviews: Climate Change*, Vol. 4, No. 1, 2013, pp. 39–60.

[7] Mia Colson, Kristin Heery, and Allan Wallis, *A Survey of Regional Planning for Climate Adaptation*, Washington, D.C.: National Association of Regional Councils, undated; and Alexander Aylett, *Progress and Challenges in the Urban Governance of Climate Change: Results of a Global Survey*, Cambridge, Mass.: Massachusetts Institute of Technology, 2014.

Climate change will present new challenges for American foreign policy in general and U.S. and global aid and development efforts in particular. Whatever the goals of the next administration, climate change will make them more difficult to achieve. Conversely, successful international action to deal with this phenomenon would represent a significant advance in the world's ability to act collectively on this and perhaps ultimately other common challenges.

Potential Disruption

Climate change has already caused observable impacts, including an increase in extreme heat events, rising sea levels, melting glaciers, extinction of species, and the disappearance of Arctic ice.[8] A vast and growing literature suggests that potential future impacts could include additional severe storms, such as those that recently struck New Orleans, New York, and the Philippines; rising sea levels that make some coastal cities difficult to inhabit; increased drought that disrupts agriculture and could spur mass migration; heat stress that increases mortality and makes it difficult to work outside in some regions of the world; significant shifts of agricultural productivity; the disappearance of Arctic ice and the attendant changes in patterns of commerce; and significant species extinctions and disruptions of the earth's ecosystems.[9]

Climate change may shift the geopolitical landscape itself, affecting the distribution of food and water and the accessibility of mineral and energy resources. The disappearance of polar ice, for instance, is opening the Arctic to commerce and to oil and mineral exploration—and conflict over those resources. Sea-level rise will undermine the viability of some ports and coastal cities while opening up opportunities for others. U.S. military operations will also be affected.[10] Bases near

---

[8] Intergovernmental Panel on Climate Change, 2013.

[9] Intergovernmental Panel on Climate Change, *Climate Change 2014: Impacts, Adaptation, and Vulnerability*, Cambridge, UK: Cambridge University Press, 2014a.

[10] U.S. Government Accountability Office, *Climate Change Adaptation: DOD Can Improve Infrastructure Planning and Processes to Better Account for Potential Impacts*, Washington, D.C., GAO-14-446, May 2014.

sea level will experience more frequent flooding or be lost entirely. Heat will affect where and when soldiers can train and fight.[11]

Over the coming decades, climate change will also reveal which countries are resilient and which are not. This may further shape the allocation of foreign assistance, particularly in the least-stable regions of the world that are, not coincidentally, least able to cope with a changing climate.[12]

Several credible estimates exist of the potential cost of climate change. The *Risky Business* project recently examined potential impacts to the United States in about a dozen sectors and projected, without efforts to adapt, economic losses that total a few percentage points of U.S. annual GDP.[13] Climate change will likely have widely disparate impacts for different populations. In particular, costs and disruptions will be higher in poor countries to adapt. If emissions keep rising, so will impacts—but not uniformly. For instance, agricultural yields may drop significantly in some regions of the world but increase in others.[14] The potential increase in global food prices between now and 2050 could range from small (a few percentage points) to large (on the order of 30 percent) depending on what one assumes about how smoothly agricultural production shifts from one region to another and the openness and efficiency of global food markets.[15] Food prices can affect economic development, public health, and political stability. Many countries place significant value on food independence; many

---

[11] Defense Science Board Task Force, *Trends and Implications for Climate Change on National and International Security*, Washington, D.C.: U.S. Department of Defense, Office of the Under Secretary of Defense for Acquisition, Technology, and Logistics, October 2011.

[12] For further information, see U.S. Agency for International Development, *USAID Climate Change and Development Strategy, 2012–2016: Clean Resilient Growth*, Washington, D.C., January 2012.

[13] M. R. Bloomberg, H. M. Paulson, and T. F. Steyer, *Risky Business: The Economic Risks of Climate Change in the United States*, Risky Business Project, 2014.

[14] Intergovernmental Panel on Climate Change, 2014a.

[15] D. Leclère, P. Havlík, S. Fuss, A. Mosnier, E. Schmid, H. Valin, M. Herrero, N. Khabarov, and M. Obersteine, "Climate Change Impacts on Agriculture, Adaptation and the Role of Uncertainty," *Impacts World 2013*, Potsdam: International Conference on Climate Change Effects, May 27–30, 2013.

communities depend on their agriculture; and some national cultures treasure their ancestral crops. Thus, the prosperity and stability of many societies may depend on their ability to adapt their agricultural systems to climate change.

To evaluate the potential extent of these systemic risks, it is useful to note that human settlements—villages and agriculture—have existed for about 10,000 years.[16] We know from evidence, such as cave paintings, that cognitively modern humans have lived on earth for much longer. But agriculture and more-complex societies appeared only at the start of the period, when the earth's climate became significantly more stable and predictable than it had generally been over the previous 100,000 years. Even during the 10,000 years of climate stability, relatively small changes, on the order of 1°C in global average temperature, have nonetheless been associated with significant changes in human society, such as the 13th century warming that allowed and then eliminated Norse settlement of Greenland and the 9th century drought that may have destroyed Mayan civilization.[17] The climate may now become significantly more variable than it has been for the entirety of human civilization. By 2050, atmospheric concentrations of greenhouse gases may be higher than they have been in about a million years. Not only was the earth's climate different in that distant past but today's rapid, fossil-fueled economic growth is compressing changes that previously took millennia into a few decades.

Twenty-first century advanced societies and economies are vastly more capable of anticipating and managing environmental stressors than were our ancestors. In optimistic scenarios, where greenhouse gas concentrations do not go much higher than double preindustrial levels, advanced societies might adapt in ways that significantly reduce the impacts of future climate change by developing drought-resistant crops and reducing vulnerabilities of coastal populations through better land

---

[16] See W. J. Burroughs, *Climate Change in Prehistory: The End of the Reign of Chaos*, Cambridge, UK: Cambridge University Press, 2005; and Ian Tattersall and Jeffrey Schwartz, *Extinct Humans*, Boulder, Colo.: Westview Press, 2001.

[17] Larry Peterson and Gerald Haug, "Climate Change and the Collapse of Mayan Civilization," *American Scientist*, Vol. 93, 2005, pp. 322–329.

use and building designs. The threat of climate change can also provide useful impetus to take actions needed for other reasons, such as creating more sustainable water systems in the American West in response to growing demand and environmental damage.

Responses

Numerous efforts to adapt to climate change are under way worldwide, but success is far from assured. To date, there have been more awareness and planning than infrastructure investments to improve climate resilience. Many of these actions taken in the United States and abroad focus on information provision and on "low-regret" strategies that have broad benefits and pose few additional costs associated solely with climate change.

Successful climate adaptation is not only about technology. Governance is also vital. It takes effective government management to tackle complex problems, such as water management and transportation planning, at a regional scale.

Cutting emissions presents a different set of political and technical challenges. Since the start of the industrial revolution, fossil fuel combustion, greenhouse gas emissions, and economic growth have all increased together. Every sector of the economy contributes to these emissions. In 2010, agriculture and land use changes contributed one-quarter of global emissions, industry one-third, buildings one-fifth, and transportation about 14 percent. These percentages, of course, vary by location. In California, for instance, transportation contributes approximately 40 percent of the state's emissions.[18]

With current technology, eliminating global greenhouse gas would require two broad changes: first, a significant increase in energy efficiency, along with elimination of the conventional combustion of fossil fuels, replaced by some combination of renewable energy and nuclear power and, second, carbon sequestration and storage. In

---

[18] California Environmental Protection Agency, Air Resources Board, "California Greenhouse Gas Emission Inventory," web page, updated December 22, 2014.

sequestration, carbon dioxide is extracted from the effluent of fossil-fuel combustion and injected underground with the expectation that it will remain trapped there indefinitely. Primitive versions of the requisite technologies exist, but to eliminate greenhouse gas emissions in the future, especially in a global economy many times larger than today's, would require deploying these technologies at orders of magnitude beyond their current scale.

It is hard to estimate with any confidence the costs of such large-scale technological transformation—perhaps akin to estimating the cost of today's telecommunications and computing in 1970s. But economists who do so suggest that the cost of stabilizing atmospheric concentrations at a relatively low level (450 parts per million, 60 percent above preindustrial levels) could cost roughly 1 percent to 4 percent of global GDP in 2030 and 3 percent to 11 percent in 2100. This does not include in the calculation various benefits, such as reduced urban air pollution or the economic benefits of avoided climate change.[19]

Challenges of Collective Action

Unlike most air pollutants whose effects are felt most where they are emitted, greenhouse gases mix evenly throughout the atmosphere, so that emissions from each country affect all others. Six jurisdictions—China, the United States, the EU, India, Russia, and Japan—account for roughly two-thirds of current emissions. Although China recently became the world's largest net carbon emitter (25 percent of global emissions, compared with 16 percent for the United States), the United States has released the most greenhouse gases over time.[20] In general, poorer countries have lower emissions per capita, so their emissions might grow significantly as they develop. For instance, the difference in emissions between China and India (25 percent versus 6 percent of

---

[19] Intergovernmental Panel on Climate Change, 2014b.

[20] World Resources Institute, "CAIT [Climate Analysis Indicators Tool] Climate Data Explorer," homepage, undated.

global emissions) parallels their difference in per capita income, which is about 4.5 times larger in China.[21]

Scientists have likened the climate change situation to a bathtub that can only hold so much water; preventing it from overflowing requires shutting off the faucet at some point. Given this cumulative nature of greenhouse gas emissions, some have begun to speak of global emissions budgets, that is, the total amount a country may emit over its entire development trajectory before transitioning to a zero-carbon economy. Since the late 18th century, humans have emitted 515 gigatons of carbon (GtC). Of this total, the United States has emitted about 96 GtC, China 41 GtC, and India 11 GtC.[22] To have even odds of limiting the global temperature increase to less than the much-discussed 2°C goal, humans can emit no more than about 1 trillion tons of carbon (1,000 GtC).[23] At current global emission rates, that trillion-ton global budget would be breached in about 50 years. The question is now to allocate the remaining budget among countries. A transition to a zero-carbon economy would shift on the order of several hundreds of billions of dollars annually away from fossil fuel extraction and power production toward investments in efficiency, renewable energy, nuclear, and carbon capture and storage.[24] The economies of some countries, including Russia and Venezuela, are much more dependent on fossil-fuel extraction than others, which also complicates collective action on limiting emissions.

---

[21] World Resources Institute, undated.

[22] See Mike Orcutt, "The United States Is Far and Away the Leader in Carbon Dioxide Emissions," *MIT Technology Review*, November 12, 2014.

[23] Myles R. Allen, David J. Frame, Chris Huntingford, Chris D. Jones, Jason A. Lowe, Malte Meinhausen, and Nicolai Meinshausen, "Warming Caused by Cumulative Carbon Emissions Towards the Trillionth Tonne," *Nature*, Vol. 458, 2009. Also see Hal Harvey, Franklin M. Orr Jr., and Clara Vondrich, "A Trillion Tons," *Daedalus*, Vol. 142, No. 1, Winter 2013.

[24] Intergovernmental Panel on Climate Change, 2014b.

Lead or Follow on Emissions Reductions

The United States may choose whether it wants to be a leader or a follower in any transition to a low-carbon economy. Reducing its own emissions increases its leverage on other countries to reduce theirs. The recent groundbreaking U.S.-China bilateral agreement on climate change is a case in point. The United States promised to cut emissions by 26–28 percent below 2005 levels over the next decade,[25] and China pledged to reduce its reliance on coal and other fossil fuels.[26]

The next administration will engage these questions in the face of decidedly mixed trends. Over the past decade, global emissions have been rising at an accelerating rate. U.S. emissions, which dropped during the Great Recession, have begun to increase again.[27] But other nations and regions are moving ahead with emission-reduction programs. For instance, Denmark and Germany now produce large percentages of their electric power from renewable sources.[28] California, Oregon, Washington state, and British Columbia have aggressive emission-reduction programs and have pledged to work to align their carbon pricing programs.[29] Emission-reducing technologies are also advancing. Developments in smart grid and distributed generation technologies worry electric utilities in the developed world about the

---

[25] The White House, Office of the Press Secretary, "Fact Sheet: U.S. Reports Its 2025 Emissions Target to the UNFCCC," Washington, D.C., March 31, 2015.

[26] The White House, Office of the Press Secretary, "Fact Sheet: U.S.-China Joint Announcement on Climate Change and Clean Energy Cooperation," Washington, D.C., November 11, 2014:

> China's target to expand total energy consumption coming from zero-emission sources to around 20 percent by 2030 is notable. It will require China to deploy an additional 800–1,000 gigawatts of nuclear, wind, solar and other zero emission generation capacity by 2030—more than all the coal-fired power plants that exist in China today and close to total current electricity generation capacity in the United States.

[27] U.S. Environmental Protection Agency, *Inventory of U.S. Greenhouse Gas Emissions and Sinks: 1990–2013*, Washington, D.C., EPA 430-4-15-004, April 15, 2015.

[28] Justin Gillis, "A Tricky Transition from Fossil Fuel: Denmark Aims for 100 Percent Renewable Energy," *New York Times*, November 14, 2014.

[29] "Pacific Coast Action Plan on Climate and Energy," San Francisco, Calif., signed October 28, 2013.

sustainability of their 20th-century business model, which is based on centralized generation and minimal storage.[30] The natural gas revolution has slowed the use of coal, among the highest-emitting fossil fuels, though it also puts price pressure on renewables, such as wind and solar.[31]

Lead

If the United States chooses to be a leader in moving toward a low-carbon economy, it would likely need to enact and expand domestic policies in three areas:

- Increase government support for research and development on low-carbon technologies for such sectors as energy, transportation, construction, and agriculture.
- Establish policies that set a price on carbon emissions through either a carbon tax or a cap-and-trade system.
- Implement complementary measures that focus on particular economic sectors, including
  - demonstration projects for such technologies as carbon sequestration
  - energy efficiency requirements in buildings
  - land use policies aimed at reducing emissions
  - renewable energy portfolio standards
  - fuel-efficiency standards in transportation
  - promoting the use of nuclear energy
  - using a social cost of carbon in estimating the cost-benefit ratios of all regulatory policies
  - financial reporting requirements for carbon-related liabilities on the balance sheets of publicly traded firms.

---

[30] Ehren Goossens and M. Chediak, "Battery-Stored Solar Power Sparks Backlash from Utilities," *Bloomberg*, October 8, 2013.

[31] U.S. Department of Energy, *Quadrennial Energy Review: Energy Transmission, Storage and Distribution Infrastructure*, Washington, D.C., April 2015.

Decarbonizing the U.S. economy by midcentury would require swift and aggressive action. Depending on its ideological leanings and political constraints, the next administration could pursue different combinations of these policy areas. For instance, a more government-guided decarbonization program could emphasize rule-making, technology demonstrations, and sector-specific policies. More market-guided decarbonization policies could emphasize basic research and development and a rising (possibly revenue-neutral) carbon tax.

An aggressive domestic decarbonization policy might increase the rate of global emissions reductions and spawn innovative technologies or industries. It could also increase goodwill toward the United States in many parts of the world.

But there are costs and risks. Rapid decarbonization could prove more expensive than current estimates. It could exacerbate regional differences within the United States, hurt regions that rely on coal and other fossil fuels (such as the Gulf Coast, Texas, and Oklahoma), and divide states and workers in winning and losing sectors of the economy. Success would depend on the rate of technological change, which would determine the savings from energy-efficiency measures and the price of low- or no-carbon alternative energy sources.[32] Research has shown that the design and implementation of government policies to abate greenhouse gas emissions matters greatly, with market-based policies that include some form of revenue recycling being most efficient.[33]

Follow

Alternately, the United States might choose to let other countries set the pace in reducing emissions. This would attenuate, or at least postpone, the difficult and costly domestic adjustments needed to reduce emissions. A wait-and-see strategy could avoid adverse economic

---

[32] McKinsey and Company, *Impact of the Financial Crisis on Carbon Economics: Version 2.1 of the Global Greenhouse Gas Abatement Cost Curve*, 2010; McKinsey and Company, *Reducing U.S. Greenhouse Gas Emissions: How Much at What Cost?* U.S. Greenhouse Gas Abatement Mapping Initiative, Executive Report, December 2007.

[33] Ian W. H. Parry and Roberton C. Williams III, *Moving U.S. Climate Policy Forward: Are Carbon Taxes the Only Good Alternative?* Washington, D.C.: Resources for the Future, Discussion Paper RFF DP 11-02, February 2011.

impacts while allowing the United States to learn from other nations' experiments in technology, practices, and regulation. It might be able to adopt these technologies at lower cost later.

Such a choice would expose the United States to the risks of damage from further rises in temperature and sea levels and of losing the innovation edge to countries that move first. Presently, the United States is the healthiest of the world's large developed economies. If it chooses not to lead, it is not clear that others will fill the void. Higher global emissions might result.

Current or New Multinational Path

At present, the United Nations Framework Convention on Climate Change (UNFCCC)—signed in 1992 and now ratified by 196 nations, including the United States—provides the foundation for international engagement on climate change. The UNFCCC calls for stabilizing concentrations of greenhouse gases at a level that "would prevent dangerous anthropogenic interference with the climate system" and established annual global meetings (the Conference of the Parties) to negotiate means to achieve that goal. The resulting 1997 Kyoto Protocol to the Framework Convention established legally binding emission reduction targets for developed countries for 2012 relative to a 1990 baseline. Signatories to the Kyoto Protocol, which did not include the United States, met their overall reduction targets, but this was largely due to a shift from coal to North Sea gas and the collapse of the energy-intensive Soviet-bloc economies after 1990.

Attempts to negotiate a successor agreement to the Kyoto Protocol have not succeeded. The 2009 Copenhagen meeting failed to produce a follow-on to Kyoto with legally binding reduction targets and timetables. Instead, it forged an agreement that the UNFCCC goal of preventing dangerous human interference requires holding global temperature increases to within 2°C of preindustrial levels.[34] However, the

---

[34] See Section 1 of the Copenhagen Accord: United Nations Framework Convention on Climate Change, "Copenhagen Accord," FCCC/CP/2009/L.7, Draft Decision -/CP.15, December 18, 2009.

voluntary emission reductions countries pledged in Copenhagen fell far short of what would be needed to meet that goal. A Paris meeting in December 2015 aims to finally achieve a binding and universal climate agreement among all the nations in the world.

Some commentators argue that the world cannot possibly achieve a consensus agreement on legally binding emission reductions consistent with the 2°C target, that no past international environmental agreement has ever committed nations to goals they do not know how to achieve, and that the economic and technological transformations needed to reach the 2°C target are large indeed. These commentators argue for a more bottom-up approach, which would de-emphasize binding targets and instead encourage bilateral and multilateral agreements to speed emission reductions, possibly by increasing research and development, sharing low-carbon technologies, and linking together emerging regional and national carbon markets (such as the European carbon-trading system and state-level programs in the United States). Such steps might be combined with voluntary national emission reductions pledges.[35]

Despite a disappointing record at best, the current approach of seeking a global consensus on a legally binding agreement still has several arguments in its favor. First is the value of building on the existing process. The current approach commands the participation of all the world's nations and the attention of many people worldwide and has a supporting institutional infrastructure. It could take many years to organize an alternative approach. Second is the value of a simple target. The 2°C goal is easy to communicate and helps focus attention, particularly for a problem like climate change that requires a vast smorgasbord of policies to address. Without a clear target to motivate action and judge performance, little might get done. Third is the importance of global justice. If powerful countries with large emissions negotiate among themselves, the vast majority of other countries, including those most vulnerable to climate change, might find their interests ignored.

---

[35] See, for instance, David G. Victor and Charles F. Kennel, "Climate Policy: Ditch the 2°C Warming Goal," *Nature*, Vol. 514, October 2, 2014; and D. G. Victor, J. House, and S. Joy, "A Madisonian Approach to Climate Policy," *Science*, Vol. 309, 2005.

The 2015 Paris negotiations may result in anything ranging from a global agreement on aggressive emissions reductions to a complete collapse of the 25-year process started by the UNFCCC. Regardless of the outcome, the next administration must decide whether the United States should continue to participate in global climate negotiations and, if so, in what role.

Pursue Current Path

Assuming that Paris negotiations produce an agreement that Congress would support, the United States could actively work within the current, albeit slightly revised, international system to make that agreement a success. Such a strategy could place the United States in more of a leadership position—with the benefits that accrue from such a role—generate goodwill abroad, potentially provide economic benefits in the transition to a low-carbon economy, and eventually stabilize greenhouse gas emissions, though likely with an average global temperature increase well above 2°C. However, even if this scenario were to come to pass, the current international framework might continue to prove ineffective, perhaps by failing to slow emissions as fast as might be possible, or collapse entirely, if the gap between aspiration and actual performance grows too large.

Set New Path

The U.S. Senate declined to ratify the 1997 Kyoto protocol. Nonetheless, the United States continued to participate in global climate negotiations but also took the lead in alternative diplomatic approaches, such as the regional bilateral agreement with China;[36] the Bush administration's multilateral technology programs;[37] and state-led efforts to establish international carbon markets, such as the Pacific Coast Collaborative among British Columbia and the four U.S. Pacific coast

---

[36] Mark Landler, "US and China Reach Climate Accord After Months of Talks," *New York Times*, November 11, 2014b.

[37] Asia-Pacific Partnership on Clean Development and Climate, homepage, undated; Major Economies Forum on Energy and Climate, homepage, 2009.

states.[38] Whether the Paris negotiations produce an agreement or not, the next administration could push aggressively to address climate change outside the framework of the UNFCCC—for instance, by continuing to pursue agreements among the small number of largest emitters or through regularizing carbon tariffs (i.e., taxes on carbon-intensive imports) and carbon-content standards for internationally traded goods as part of the WTO regime.[39] Such actions might prove more effective in limiting climate change than the current UNFCCC structure but might hasten the collapse of that structure and increase hostility to the United States among some nations, with possible spill-over effects into security and other foreign policy matters.

---

[38] The four states are Alaska, California, Oregon, and Washington. See Pacific Coast Collaborative, homepage, undated.

[39] Also see William Nordhaus's proposal for a "Climate Club," in which countries that established a domestic carbon price at some agreed-to level would be allowed to impose a tariff on all countries that had not established such a domestic carbon price (William D. Nordhaus, "A New Solution: The Climate Club," *The New York Review of Books,* June 4, 2015).

CHAPTER SEVEN

Europe

Russia's annexation of Crimea and its invasion of eastern Ukraine have sent shockwaves throughout the West. These events call into question two key assumptions on which American policy in Europe has been based since the end of the Cold War: (1) that Europe was stable and secure and that, as a consequence, the United States could focus more attention on other regions, particularly the Middle East and Asia and (2) that Russia had become more of a partner than an adversary and had a common interest in building a "Europe whole and free and at peace."[1]

Russia's actions in Crimea and eastern Ukraine suggest that these assumptions no longer provide a sound basis on which to base U.S. policy. The next administration will have to adopt a new approach toward a more assertive and less predictable Russia. Yet, despite the Russian challenge, European defense budgets remain at an all-time low as a proportion of GDP, and many are still going down, as indeed is that of the United States (see Chapter Three).

The Ukraine crisis comes at a time when European institutions are under mounting economic and political pressures. Several major economies may be facing renewed recession. Greece may pull out of the euro, and the United Kingdom is threatening to pull out of the EU. Far-right parties, some of which actually find inspiration in Putin's Russia and all of which challenge important elements of the European and transatlantic construction, are showing growing strength at the

---

[1] George H. W. Bush, "Remarks to the Citizens in Mainz," May 31, 1989.

polls. Homegrown terrorism is highlighting the problems of integrating large and still-growing immigrant populations.

Confronting or Engaging Russia

The Russian annexation of Crimea and invasion of eastern Ukraine threatens the post–Cold War order in Europe and sets a dangerous example for other would-be regional hegemons, most notably China. To discourage both repetition and imitation, it is important to ensure that this aggression is seen not to pay. At the same time, it is not in the American or European interest to intensify or expand the conflict in Ukraine, and any resolution of the crisis will require some accommodation to Russia's self-perceived interests. Russian cooperation is also important in other areas, most notably Iran; Afghanistan; and, potentially, Syria. U.S. policy will need, therefore, to establish some balance among the needs to punish Russia for prior bad behavior, to deter any future such behavior, to deescalate the actual fighting in Ukraine, to resolve that conflict if possible, and to cooperate with Russia on other areas of possible common interest. U.S. policy will need to combine, in some degree, confrontation with engagement, the exact balance to be determined depending in large measure on future Russian behavior.

Confrontation

The next administration could seek to further increase the economic, political, and military pressure on Russia in an attempt to compel Moscow to adopt a more cooperative and conciliatory policy. These steps would be designed to increase the costs to Moscow of attempts to coerce Kiev into abandoning its pro-European course. The administration could significantly step up military assistance—including lethal military assistance—to Ukraine in an attempt to enhance Kiev's ability to defend what is left of its territory or even to recapture what it has lost in eastern Ukraine. In line with this course, Washington could also decide to actively support Ukraine's eventual entry into NATO and the EU.

The United States could also move to bolster the defense of exposed NATO nations, particularly the Baltic states, beyond the rotational deployments of allied forces already agreed on. (Possibilities for so doing are explored further in the section on Europe in Chapter Three.)

Taking a tough line with Russia may reassure other allies and partners regarding the dependability of U.S. support. On the other hand, Russia will always care more about Ukraine than the United States, and Putin can probably overmatch any U.S. commitment, leading perhaps to an even greater impression of American failure. Putting U.S. and western European troops directly on Russia's borders could strengthen deterrence while reassuring the allies least able to ensure their own defense. It might also lead Russia to expand its own efforts at subversion and intimidation.

Engagement

To date, Washington has left mediation in the Ukraine conflict largely to the Europeans, most notably to Chancellor Angela Merkel of Germany. Washington has supported the cessation of most NATO-Russia activities and shut down military-to-military contacts, while preserving its own bilateral dialogue on a few issues of common interest. The United States has led and even gotten slightly out in front of Europe in imposing economic sanctions. In the future, the United States could become more active in seeking to resolve the Ukraine crisis, as described in the following section. To the extent that this effort made any headway, Washington could begin scaling back sanctions, advocating renewed NATO-Russia ties, and reintroducing Russia into the G-8 consultations.

Engagement with Russia keeps lines of communication open, offers some prospect of eventual peaceful resolution of the Ukraine conflict, and keeps some areas of current cooperation on track. It can also foster the impression that the United States cares little about Ukraine's future and is an undependable ally and partner. Too passive a stance on Ukraine could encourage China to believe that the United States would not respond significantly to forceful assertion of its maritime claims in East Asia. And failing to strengthen NATO's defense posture on its eastern flank would run the risk that Putin will take the opportunity to seize more territory in a quick invasion, confronting NATO with a fait accompli.

Whither Ukraine: Partition or Neutral Buffer

Current U.S. policy is that Ukraine should be united and free to choose whether to seek EU and NATO membership. There seems to be no possibility that Russia would allow both these objectives to be achieved. Pursuing both probably ensures that neither will be accomplished—that is, Ukraine will remain divided and will not progress toward either EU or NATO membership. This outcome, which means effectively consolidating another frozen conflict on Russia's borders, may well meet minimal Russian, American, and European needs but is also the least-attractive outcome for Ukrainians. It is therefore worth considering two other alternatives.

Partition

Ukraine, shorn of the areas now controlled by Russia, has expressed a clear desire to move toward both EU and NATO membership, and to undertake the reforms necessary to become eligible. The United States could support these aspirations, not just rhetorically, but practically, by improving Ukraine's capacity to defend its remaining territory and by urging the initiation of a NATO membership action plan. Moving in this direction would almost certainly end any prospect of Ukraine regaining control over the breakaway areas in its east (not to speak of Crimea, which is in any case probably lost for good). Ukraine might not have to renounce its claim to these territories to secure EU and NATO membership, but it would certainly have to renounce any intention to use force to make good on that claim. Movement in this direction might stimulate renewed Russian aggression—leading to a more open military conflict—and expanded Russian control of additional Ukrainian territory, although it is unlikely that Russia would or could seize the country as a whole. Wherever the line would ultimately be drawn, most of Ukraine could, assuming significant internal reforms take place, become part of the West and the rest effectively part of Russia.

Neutralization

Rather than steering Ukrainian leaders toward a Western alignment that almost certainly involves heightened violence and the continued division of their country, Washington might encourage and help mediate a settlement of the crisis that produces a reunified Ukraine (sans Crimea) aligned neither with Russia nor NATO. Ukraine could continue to maintain its claim to Crimea, and Western sanctions imposed after the Russian seizure of Crimea but before its aggression in eastern Ukraine could remain in force. Ukraine would secure control of its borders. Some degree of autonomy, particularly in cultural areas, would be provided to eastern Ukraine. Ukraine would forswear NATO membership but continue to pursue closer ties with the EU. Other than the Crimea-specific sanctions, normal U.S., NATO, and EU ties with Russia could be reestablished.

Either of these solutions might stabilize Ukraine. One requires an implicit acquiescence in Russian aggression, setting a bad example for it and others. The other alternative sacrifices a point of principle, that countries should be free to choose their geopolitical alignment regardless of the views of their more powerful neighbors.

Addressing European Disintegration

The challenges posed by a more assertive, less predictable Russia come at a time when much of Europe is economically weaker and politically more fragmented than any time in the recent past. There are signs that the crisis in Ukraine and growing tensions with Russia may be combining with other factors to push the Eurozone back into recession.[2] During the first quarter of 2014, the economies of seven out of 18 Eurozone countries shrank while France had zero growth.

The escalating sanctions imposed by the United States and Europe have sent jitters through European financial circles. Leading figures in the European financial world, such as Mario Draghi, president of the

[2] Jack Ewing and Gaia Pianigiana, "Italy Falls Back into Recession, Raising Concern for Eurozone Economy," *New York Times*, August 6, 2014.

European Central Bank, fear that a further breakdown of relations with Russia over Ukraine could impede European growth. This may make European political leaders more reluctant in the future to impose additional sanctions on Russia and could generate increasing pressure to roll back some of the restrictions already imposed on Russia.

Moreover, politically, the EU faces a major challenge from the United Kingdom, where anti-EU sentiment has visibly increased. The United Kingdom has threatened to withdraw from the EU if its demands for greater legal autonomy are not met. If it were to withdraw, this could pose a grave threat to the EU. Allowing Greece to leave the euro is one thing, but a British decision to quit the EU itself quite another. A withdrawal by the United Kingdom might incite other members to question their own commitment, leading to a major weakening of the entire edifice of the EU.

Bolster Europe, Oppose Fragmentation

Historically, the United States has supported continued advances in European integration, even in cases where Washington had doubts about the feasibility or desirability of certain initiatives, most notably the creation of a common currency and the move toward common foreign and security policies. The result has been a generally stable, peaceful, cooperative continent and a prosperous trading partner. Consistent with this long-term approach, the United States might look for further ways to help Europe overcome its current malaise. Principal among these would be conclusion of a transatlantic free trade area (i.e., TTIP), which would stimulate both European and American growth and buttress long-standing transatlantic security ties. Washington might also strengthen macroeconomic coordination with the Eurozone, more openly express concern about a UK withdrawal from the EU, and become more vocally critical regarding some of the more extreme (and bizarrely pro-Russian) political movements on Europe's populist right.

Leave Europe to the Europeans

Overt U.S. engagement in Europe's internal debates might be judged counterproductive, actually influencing opinion in the wrong direction. Washington might prefer to concentrate on first securing the transpacific free trade area (i.e., TPP), leaving the transatlantic initiative to a later date. Some of the weaker European economies might actually achieve higher growth outside the Eurozone. A United Kingdom outside the EU might be an even more willing partner for the United States. For all these reasons, Washington might choose to remain largely passive as Europe confronts its current political and economic travails.

CHAPTER EIGHT

East Asia

Over eight presidencies, Washington has sought to build a stable and cooperative relationship with China, one that would enable the two countries to work together on issues of mutual concern—most recently nonproliferation, terrorism, cybersecurity, economic cooperation, and climate change—while allowing them to manage their differences constructively.[1]

China's growing economic and military power, involvement in cyber theft of intellectual property, refusal to renounce the use of force as a means of resolving the status of Taiwan, and assertive advancement of territorial claims affecting several of its neighbors, including U.S. allies Japan and the Philippines, presents the United States with serious challenges. The United States does not take any position on the sovereignty issues at the heart of the maritime territorial disputes between China and other countries in the region, but Washington aims to deter China from using intimidation, coercion, or force to settle these differences.[2] U.S. defense treaties with Japan and the Philippines mean the United States could become involved in a conflict with China if one of these disputes escalates, as might be the case in a military confrontation between China and Japan or China and Taiwan.

East Asia is also the world's fastest-growing region, offering the largest opportunity for American exporters and investors.

---

1 See, for example, John Kerry, "Remarks on U.S.-China Relations," Johns Hopkins School of Advanced International Studies, Washington, D.C., November 4, 2014.

2 See, for example, Chuck Hagel, "IISS Shangri-La Dialogue," speech, Singapore, May 31, 2014.

In recognition of these trends, the Obama administration signaled its intention to pivot, or rebalance, American attention and security resources toward East Asia.

China perceives that the United States intends to contain China to prevent its growing power and influence from challenging U.S. interests.[3] At the same time, some Chinese analysts question whether Washington will be able to muster the will and resources to maintain its traditional position in the region in the face of its domestic political challenges, budget problems, and crises in the Middle East and Europe. Some in the United States fear that China is intent on reshaping important elements of the current international order to increase its influence and diminish that of the United States.

Central to Washington's intention to pivot to East Asia has been the effort to modernize U.S. alliances and strengthen diplomatic, economic, and security relationships with partners throughout the region. In doing so, the United States has encountered growing demand for assurance from allies and partners who are wary of China's growing power and more assertive foreign policy. Countries in the region are looking for signs that the pivot will be substantive and sustainable, despite U.S. budgetary constraints and Washington's need to deal with mounting security problems in other parts of the world.

India, long committed to maintaining a position of flexible neutrality in international affairs, has now undertaken efforts to build closer ties to regional powers Japan and Australia, and even the United States. Further moves by China to expand its military presence in the Indian Ocean or apply pressure in the disputed border areas between the two countries could cause India to align itself more closely with the United States and its allies, something previous Indian governments have always resisted.

The pivot (later rebranded a rebalancing) to East Asia was announced shortly after the U.S. military withdrawal from Iraq and

---

[3] On Chinese concerns about containment, see Ely Ratner, "Rebalancing to Asia with an Insecure China," *Washington Quarterly*, Spring 2013; Andrew J. Nathan and Andrew Scobell, "How China Sees America: The Sum of Beijing's Fears," *Foreign Affairs*, September/October 2012; and Michael S. Chase, "Chinese Suspicion and U.S. Intentions," *Survival*, June–July 2011, pp. 133–150.

at a time when the United States was also beginning to disengage from Afghanistan. Further reductions in the U.S. military presence in Europe were also planned. Subsequently, serious new challenges have arisen in both Europe and the Middle East, making it hard to sustain further shifts in U.S. military forces or high-level attention. U.S. forces have drawn down considerably in Afghanistan, though Obama has slowed the withdrawal, but the U.S. troop presence in the Gulf and Iraq is increasing, and any further reductions in Europe are unlikely to go forward. While East Asia will continue to receive considerable American attention, the shift of resources and attention from other regions is probably over.

China remains an essential economic partner and shares the interest with the United States of restraining North Korea. Pyongyang does not appear to want to start a large-scale conflict with South Korea and the United States, but its development of nuclear weapons and long-range missiles threatens the United States and its neighbors, and its proliferation activities fuel conflicts in other regions. Its often provocative and unpredictable actions hold the risk of uncontrolled escalation.

Accommodating Versus Containing China

While Chinese economic growth is slowing, it will likely continue to exceed that of the United States and most of its East Asian neighbors. The United States has sought to counterbalance growing Chinese military, economic, and political power in East Asia by strengthening ties to most of the other states in the region. It has sought to channel growing Chinese influence within international institutions in ways that preserve and strengthen the current institutional order. China, for its part, is beginning to seek regional and global influence commensurate with its strength. China does not wish to overturn the international order, from which it profits greatly, but rather to reshape it in ways that reflect its growing power and to establish over time its position as the regional hegemon.

Across any number of spheres, the United States faces choices about whether to counter and contain Chinese influence, collaborate with it for common purposes, or accommodate it in ways that minimize damage to U.S. interests.

Accommodation

The objective of an accommodation approach would be to find ways to allow China to develop influence and responsibilities commensurate with its power within, rather than outside, the current normative and institutional order. Proponents of this view prefer to focus on building trust, and some have advocated reducing or eliminating activities that Beijing sees as associated with a presumed U.S. intent to contain China, such as by reducing or ending U.S. reconnaissance activity off of China's coast or by curtailing or eliminating arms sales to Taiwan. More generally, an approach that focuses largely on engagement might involve more fully exploring China's vision of "new type of major country relationship."[4] This might help assuage Chinese concerns about U.S. containment and address problems related to mutual suspicion in the relationship.[5]

While China is not seeking to overthrow existing multilateral regimes, it is beginning to construct and lead supplementary institutions and arrangements for shaping the international order. Rather than opposing such efforts, the United States might offer to support and participate where appropriate in such Chinese initiatives. The United States might invite China to become a member of TPP trade agreement.

An approach that focuses so heavily on engagement and accommodation would expose the United States to a number of risks. It would result in a loss of influence in this dynamic region, making it more difficult for the United States to defend its interests there. Such an approach could intensify the concerns of U.S. allies and partners who already worry that Beijing seeks to persuade the United States to sacrifice their interests in pursuit of closer ties with China. The

---

4 "Xinhua Insight: China Marks Six Priorities for New Type of Major-Country Relations with U.S.," *Xinhua*, November 12, 2014.

5 On mutual suspicion, see Kenneth Liberthal and Wang Jisi, *Addressing U.S.-China Strategic Distrust*, Washington, D.C.: Brookings Institution, March 2012.

intensification of such concerns could motivate some countries to take matters into their own hands in ways that could prove destabilizing. Indeed, regional perceptions of a disengaged or acquiescent Washington would only intensify fears and insecurity in regional capitals and could very well increase the likelihood of an arms race, miscalculation, and conflict. Finally, an approach that largely eschews balancing would leave the United States poorly prepared for the possibility of a more aggressive turn by China. And it might unintentionally encourage more-aggressive behavior if China perceived it as stemming from U.S. weakness, distraction, or unwillingness to maintain its long-standing regional security commitments.

Containment

Alternatively, the United States could place greater emphasis on counterbalancing Chinese power and limiting the grown of its influence. A number of analysts have advocated different variations of such a strategy.[6] A stronger focus on balancing could do more to deter China from using force or the threat of force to resolve maritime territorial disputes on its own terms. It might also help assure U.S. allies that Washington will put their interests first. Some specific initiatives the United States might consider as part of such an approach could include deployment of larger numbers of U.S. military forces in the region; accelerated modernization of key U.S. military capabilities and operational concepts; increased arms sales to allies and partners; and closer defense cooperation with emerging partners, most notably Vietnam, which is currently seeking to bolster its ability to push back against China's more assertive pursuit of its claims in the South China Sea. Yet shifting to a strategy that is more heavily focused on counterbalancing would have a number of costs. Additional U.S. forces for the region, an enhanced basing posture, and accelerated force modernization would either come at the expense of other regions or require a significant increase in the U.S. defense budget. A more confrontational approach toward Beijing would fuel Chinese suspicion that Washington's intentions are hostile

---

6 Aaron Friedberg, "Bucking Beijing: An Alternative U.S. China Policy," *Foreign Affairs*, September/October 2012.

and that it seeks to encircle China. This would risk sacrificing opportunities for cooperation on climate change, North Korea, Iran, and other issues of importance to the both countries. Second, allies and partners and other countries in the region want to avoid being put in a position where they need to choose between the United States and China,[7] and a more adversarial U.S.-China relationship could make them feel that they have to make difficult decisions about their defenses and political orientation in the region. Many countries would also worry that a more confrontational U.S.-China relationship would be bad for regional stability and thus damaging to their economic and security interests. A more adversarial approach might result in a level of military competition that is destabilizing and potentially unaffordable. Finally, too unqualified a commitment to regional allies could find the United States embroiled in a serious conflict over some territorial dispute of little inherent concern or importance to American interests.

(Chapter Two contains additional detail on the options for containing or accommodating growing Chinese influence in the trade and finance fields. Chapter Three expands on the additional U.S. military capabilities that may be needed to counterbalance growing Chinese capacity for power projection.)

Isolating or Engaging North Korea

Pyongyang's military provocations and its nuclear and missile programs have long been serious threats to regional stability and to the security interests of the United States and its allies. Following a series of nuclear tests and missile launches in recent years, Pyongyang has threatened to conduct another nuclear test and has displayed road-mobile ICBMs in recent military parades.[8] North Korea is extremely unlikely to bargain

[7] Evan S. Madeiros, Keith Crane, Eric Heginbotham, Norman D. Levin, Julia F. Lowell, Angel Rabasa, and Somi Seong, *Pacific Currents: The Responses of U.S. Allies and Security Partners in East Asia to China's Rise*, Santa Monica, Calif.: RAND Corporation, MG-736-AF, 2008.

[8] Office of the Secretary of Defense, *Military and Security Developments Involving the Democratic People's Republic of Korea*, Washington, D.C.: U.S. Department of Defense, 2014.

away its nuclear and missile programs, which it regards as essential for regime survival, strategic deterrence, and generating bargaining leverage to coerce concessions and assistance from the outside world.[9] North Korea also uses computer network operations against South Korea, a tactic it has been employing for several years against its rival, and more recently against Sony Pictures, a U.S.-based commercial enterprise.[10]

The United States has approached the North Korean challenge by working closely with its allies in South Korea and Japan and by deploying conventional U.S. forces and nuclear capabilities for extended deterrence. The United States would like to find ways to encourage South Korea and Japan to cooperate more closely but faces constraints because of difficult relations between Tokyo and Seoul.

Washington has also sought to gain China's assistance in countering the North Korean nuclear problem. China is increasingly frustrated with Pyongyang's provocative and unpredictable behavior, even as it is reluctant to work overtly with the United States. In theory, Beijing shares a strong common interest with Washington and Seoul in bringing about positive change in the north but is extremely reluctant to do anything it sees as risking further destabilization of the economic or political situation in its longtime ally in Pyongyang.[11]

At different times, the two main U.S. allies in the region, South Korea and Japan, have pursued divergent strategies toward North Korea. South Korea has sought to soften North Korean hostility via engagement and even limited forms of cooperation, while Japan has preferred a harder line of effort to isolate and penalize North Korea.

Isolation and Containment

An approach centered on isolation, containment, and counterproliferation could be sufficient to slow North Korea's nuclear and missile development and constrain Pyongyang from exporting nuclear and missile

---

[9] Jonathan Pollack, *No Exit: North Korea, Nuclear Weapons, and International Security*, London: International Institute for Strategic Studies, May 2011.

[10] Office of the Secretary of Defense, 2014, p. 11.

[11] Andrew Scobell and Mark Cozad, "China's North Korea Policy: Rethink or Recharge?" *Parameters*, Vol. 44, No. 1, Spring 2014.

technology to countries in other parts of the world but is unlikely to result in the denuclearization of North Korea. Even with the continuation of economic sanctions, the North Korean regime would likely retain formidable capabilities to threaten U.S. interests and those of U.S. allies, especially South Korea and Japan.

The United States will continue working with allies and partners to curtail North Korean nuclear and missile development; deter provocative actions, such as North Korea's sinking of the South Korean ship *Cheonan* and its shelling of Yeonpyeong Island in 2010; and prevent nuclear and missile technology proliferation. In addition, the United States might need to strengthen regional and national missile defense capabilities as Pyongyang moves forward with such capabilities as its mobile ICBM program but would need to do so in a way that does not undermine stability in the U.S.-China relationship, given Beijing's suspicion that the missile defense systems the United States is deploying to deal with North Korea are also intended to undercut the credibility of China's nuclear deterrent. This balance should be manageable, however, given that China is modernizing and increasing the size of its nuclear force and that North Korea's progress has been relatively slow in some key areas. Notably, Pyongyang has not flight-tested its Hwasong-13 road-mobile ICBM, which means its reliability would be low.[12]

As noted, this approach might slow down, but not halt—let alone reverse—improvements in North Korea's nuclear and missile capabilities. The absence of regular contacts would leave each side less informed about the other's concerns and intentions. It might make North Korean attention-seeking provocations more likely and might increase the risk of miscalculation in any resultant crisis.

Another possible consequence might be a reduction in China's willingness to cooperate with the United States on North Korea if Beijing perceives the U.S. approach as likely to result in instability or to heighten risks of crisis or conflict on the Korean Peninsula.

---

[12] Office of the Secretary of Defense, 2014, p. 10.

Engagement and Limited Cooperation

An alternative to isolation and containment is more frequent and direct bilateral contact with North Korea. This could include some mix of sanctions relief, multiparty talks, and direct dialogue between North Korea, the United States, and U.S. allies.[13] The objective would be to secure more temperate and predictable North Korean behavior, reducing provocations and the danger of miscalculation leading to war.

It is difficult to judge North Korean intentions with any certainty, but there have been at least some indications that Pyongyang would be receptive to such an approach. Indeed, at least prior to North Korea's recent cyberattack on Sony Pictures, Pyongyang appeared to be showing signs of interest in renewed engagement with the United States, South Korea, and other countries.[14]

Even if North Korea proved receptive to more-extensive contacts, these might do little to eliminate Pyongyang's military threats or alter its provocative behavior. Indeed, some observers have assessed that previous attempts to engage with North Korea have had a relatively poor track record in changing its threatening behavior. Even China, North Korea's most important source of economic and diplomatic support and thus the country that should have the greatest influence in dealing with Pyongyang, has had little success in changing North Korean behavior.[15] Pyongyang will almost certainly remain unwilling to give up its nuclear and missile capabilities under any circumstances, no matter what assistance or assurances Washington, Beijing, Seoul, and Tokyo may be willing to provide.

---

[13] See, for example, *North Korea Inside Out: The Case for Economic Engagement*, Report of an Independent Task Force convened by the Asia Society's Center on U.S.-China Relations and the University of California Institute on Global Conflict and Cooperation, December 2009.

[14] Mark Landler, "Latest North Korean Mystery: A Diplomatic Charm Offensive," *New York Times*, October 23, 2014a.

[15] See, for example, Richard Weitz, "North Korea: The Problem with Reconciliation via Engagement," *The Diplomat*, January 23, 2014; and Mathieu Duchatel, "Tactical Pause in China's Economic Engagement with North Korea," *China Brief*, Vol. 14, No. 9, May 7, 2014.

CHAPTER NINE

Middle East and South Asia

By the end of 2016, the Middle East will be 15 years removed from 9/11 and the invasion of Afghanistan and 13 from the U.S. invasion of Iraq, as well as five years removed from a wave of uprisings that toppled long-standing autocrats in Tunis, Cairo, Tripoli, and Sana'a; led to a bloody civil war in Syria; spooked Arab Gulf allies; spawned a counter-revolution in Egypt; and left a political vacuum in Libya and Yemen. Throughout this period, the United States has pursued a number of core interests in the Middle East and South Asia, including maintaining the free flow of oil and gas out of the Persian Gulf; bolstering the security of U.S. partners and allies, particularly Israel, Jordan, and the Gulf Cooperation Council states; and mitigating the threat of violent extremism within and beyond this region.

The Israeli-Palestinian conflict, the Iranian Revolution, 9/11, the wars in Afghanistan and Iraq, and the emergence of ISIS have all made deep impressions on the American public, making this region particularly salient in U.S. policy and politics for much of the past quarter century. This will continue because the conflicts in Afghanistan, Pakistan, Iraq, Syria, Libya, and Yemen are likely to persist for years to come. The regimes not yet themselves confronting local insurgencies will endeavor to keep the lid on popular pressures and work to suppress militant threats at home and in neighboring states.

By the end of 2016, ISIS may have been driven underground throughout much of Iraq, but it will remain a major force in Syria. Even if the P5+1 (the five permanent members of the United Nations Security Council plus Germany) negotiations with Iran produce a final agreement limiting Iran's nuclear program, that settlement will remain highly controversial in both the United States and Iran, as well as throughout the region, uniting in this respect both Israel and many Sunni governments. The war in Afghanistan will continue, and violent militant groups will pose a serious threat to nuclear-armed Pakistan.

Engaging or Confronting Iran

The United States and Iran are engaged in multiparty talks regarding the future of Iran's nuclear program. The conclusion of a final nuclear accord with Iran would open the possibility of further rapprochement, leading even to a renewal of diplomatic relations between the two countries. These possibilities will be limited by substantial domestic resistance in both countries and strong countervailing pressures on the U.S. administration from Israel and the Gulf states. The next U.S. administration could work to isolate Iran in the hopes of weakening the regime or could build on areas of mutual U.S.-Iranian interest with an eye toward establishing basic rules of the road over how the U.S. and Iran interact in the region.

Engagement

Iranian and U.S. policies are substantially aligned in both Afghanistan and Iraq, in that both countries support the same local regimes and oppose the same adversaries. Iraq nevertheless supports violent and highly sectarian elements within Iraq that threaten that country's unity. U.S. and Iranian policies toward Syria are aligned in one respect and sharply divergent in another, in that Iran supports the regime of President Bashar al-Assad, while the United States opposes it, but both Iran and the United States oppose ISIS. Successful conclusion of a nuclear agreement would make overt cooperation easier where the two sides' policies largely coincide and would also permit the two governments to explore ways to close the gap where large differences remain. More fully aligning U.S. and Iranian policies regarding Syria is probably essential to ending that civil war. Engagement of this sort could bring significant benefits, in the form of enhanced U.S.-Iranian cooperation in checking Sunni jihadists and creating a pathway for a negotiated solution in Syria, albeit one that would likely require significant continuity in the composition of the regime.

Confrontation

In the absence of a nuclear accord, continued U.S.-Iran confrontation is inevitable. Indeed, the next administration might choose to take direct military action to set back Iranian nuclear weapon development. Even with an accord, Iran is unlikely to abandon its support for Hezbollah and Hamas or for Shi'a minorities in other regional states, thus putting it at continuing odds with Israel, Saudi Arabia, and most of the other Gulf states. Suspicions about Iran's adherence to any accord limiting its nuclear program would persist, perhaps even leading the next administration to walk away from any agreement reached under the current American leadership. Powerful elements within the Iranian regime will, in any case, seek to weaken the U.S. role as security provider to the Gulf Cooperation Council states. Taking a tough line with Iran would bolster most other U.S. relationships in the region but would only have the desired restraining effect on Iran if the United States were able to hold together the current broad international consensus favoring sanctions and other forms of pressure. Iran could respond to intensified American pressures by undermining U.S. efforts and targeting U.S. personnel in Iraq and Afghanistan.

Ending the War in Syria

The combined weight of Kurdish, Shi'a, moderate Sunni, Iranian, U.S., and other regional military efforts may eventually succeed in driving ISIS out of much of Iraq, or at least underground. As long as the civil war in Syria continues, however, it will attract and radicalize young Muslims from all over the world, including the United States. It will also contribute to heightened hostility between Shi'a and Sunni and between Arabs and Persians, thereby contributing to other conflicts in the region. The choice the next administration will face is whether to prioritize defeating ISIS or overthrowing Assad, recognizing that weakening either tends to strengthen the other.

ISIS First, then Assad

The United States could commit itself to the sequential defeat of both opponents, first ISIS and then Assad. To succeed on both scores, this approach would require a major commitment to raising, training, equipping, and supporting moderate alternatives to both Assad and ISIS. It would almost certainly require an eventual U.S. air campaign against the Syrian regime and possibly the eventual introduction of U.S. ground forces. America's regional allies will favor such an approach. Iran and Russia, both deeply engaged in supporting the current regime, will do their best to frustrate its second, anti-Assad phase. Support from elsewhere in the international community is likely to weaken once ISIS has been marginalized.

Target ISIS, Promote a Peace Settlement

Under this alternative, the United States would step up targeting of ISIS and other extremist elements in Syria while seeking to work with Russia, Iran, and U.S. allies in the region on a settlement that combines significant elements of the Assad regime with moderate opposition groups. This arrangement might require some transitional role for Assad himself. Any such deal would be highly controversial domestically, internationally, regionally, and within Syria. Such a settlement does, however, offer the prospect of bringing the Syrian civil war to a close earlier than would the ISIS in Iraq first, then Assad alternative, albeit with a less satisfactory end state.

Arab-Israeli Settlement

There are, theoretically at least, three alternative outcomes to any Israeli-Palestinian peace process: a two-state solution, a one-state solution, or the status quo, which is to say no solution. So far, most Palestinians (and other Arab governments) and Israelis have preferred the status quo to any alternative outcome that the other side would accept. Moderate Arabs and Israelis support, in principle, a Jewish state and an Arab state living side by side in the former Mandate of Palestine. More-radical elements on both sides favor a one-state solution, covering all or most of historic Palestine, although they obviously differ on whether Arabs or Jews would dominate the resultant state.

Status Quo

The next administration may choose to continue expressing support for the two-state solution and may even seek to mediate negotiations toward that end, while leaving it to the parties to define a mutually acceptable outcome. This approach has been tried for 40 years without success. The probable failure of another such effort will be extension of the status quo—that is, the Israeli occupation of the West Bank and blockade of Gaza, the continued expansion of Israeli settlements in occupied territories, and limited autonomy for Palestinian authorities within noncontiguous areas not populated or fully controlled by Israelis. This approach enjoys bipartisan support in the United States and Israel and quiet acquiescence from most Arab governments. It allows the Israelis to postpone the difficult choice between the democratic and Jewish characters of their state. Perpetuation of the status quo also promotes continued resistance among Palestinians, feeds the radical narrative throughout the rest of the Middle East, and will likely make Israel even more dependent on U.S. support in the United Nations and other forums as it continues to lose sympathy in Europe.

Two States

Alternatively, the next administration might seek to take American efforts to secure a two-state outcome to a new level. The United States might itself lay out the parameters of an equitable peace settlement and have them endorsed by the United Nations Security Council. It might cease vetoing resolutions of the Security Council that are fully consistent with U.S. policy but opposed by Israel. It might reduce levels of economic support for Israel, including financing of its military purchases. At the same time, the United States could seek European and Arab government support in putting comparable pressures on the Palestinian side. Such steps would be applauded internationally and regionally, but would provoke strong domestic resistance in the United States. By raising the price for Israel and the Palestinians to maintain the status quo, such a policy might eventually incline the two toward a more serious effort to conclude a two-state solution. Alternatively, it might accelerate the trend in Israel toward formal incorporation of some or all of the West Bank.

One State

Hard as it is to envisage a two-state outcome that both sides could accept, it is even harder to imagine how the existing populations of Israel, the West Bank, and Gaza could live together in one state. Palestinian proponents of such a solution assume the return of Palestinian refugees from abroad and a consequently large Arab majority in such a state. Israeli advocates of "Greater Israel" assume the opposite, that continued Jewish immigration would be encouraged and any significant Arab immigration forbidden. It is possible to imagine formulas that would preserve the current ethnic balance in the combined territory and would therefore yield a narrow Jewish majority, but it is hard to imagine either side being satisfied with such an arrangement.

Nevertheless, if the next administration were to conclude that perpetuating the status quo for another eight years was unacceptable, it might begin speaking of the one-state solution not as its preferred outcome but as one more acceptable than no solution at all. As a practical matter, this would mean taking Netanyahu's rejection of a two-state solution at face value and beginning instead to press for civil and political rights for the Palestinian population of the West Bank and Gaza. Such a stance by Washington might prove the most effective way of securing stronger Israeli support for a two-state outcome. Indeed, beginning to talk seriously and positively about a one-state solution is probably better viewed as a tactical move to drive the parties toward the two-state model than itself an achievable goal. Like the two-state alternative, this approach would be quite controversial domestically.

Afghanistan and Pakistan: Staying or Going

President Obama has committed to withdraw all U.S. troops from Afghanistan by the end of 2016. Afghan President Mohammad Ashraf Ghani has already asked Obama to reconsider this move. In any case, whether the United States retains a military presence in Afghanistan in 2017 and beyond is ultimately an issue for Obama's successor and is an issue likely to arise in the course of the 2016 presidential campaign. Much will obviously depend on how the war with the Taliban goes in the interim. Most likely, the Taliban will gain ground in the countryside over the next two fighting seasons but fail to take any of the major cities. Pakistan may begin to put some pressure on Afghan Taliban sanctuaries within its border regions, but Pakistani forces will have their hands full dealing with their own indigenous insurgency.

Stay

The United States might chose to retain a military training, advisory, and assistance force in Afghanistan. This force would not routinely engage in ground combat but would conduct occasional CT operations against al Qaeda and its supporters and imitators. Within its limited capabilities, this force might also provide intelligence and close air support for Afghan units. It would also cooperate with both Pakistan and Afghanistan in combating violent militant groups on both sides of that border. This alternative might involve a deployment of 5,000 to 10,000 U.S. troops, perhaps with a smaller number from other NATO countries. This commitment would not be sufficient to defeat, or even greatly weaken, the Taliban but could deny it control over most of the population, sustain broader regional and international support for the Kabul government, help Pakistan deal with its own militant threat, and boost the chances for an eventual Afghan peace process.

Go

Alternatively, the United States might continue its diplomatic and financial support for the Afghan government, including funding and arming its security forces, without any deployed military presence beyond those DoD personnel attached to the U.S. embassy. This would bring to an end a 15-year military engagement in Afghanistan, save the American taxpayer at least $5 billion to $10 billion per year, and increase the level of U.S. military forces available for other contingencies. It might demoralize the Afghan government and security forces, exacerbate divisions within the society, and cause regional powers to hedge their support for the Kabul regime. India and Pakistan, in particular, would likely begin jockeying to fill the vacuum left by the U.S. withdrawal. Extremist groups seeking to overthrow the Pakistani regime would find it easier to operate out of sanctuaries on the Afghan side of the border. Pakistan would likely expand its own support for the Afghan Taliban. The result might be a cascading series of defeats for Afghan forces, perhaps leading to the collapse of the elected government and a much wider and more violent civil war. Finally, the complete withdrawal of U.S. forces would pretty much eliminate American ability to deal with residual al Qaeda or other terrorist threats to the United States homeland originating from either side of the Afghanistan-Pakistan border.

CHAPTER TEN

Rethinking American National Strategy

The Obama administration put in place a national strategy of retrenchment—in Europe, the Middle East, and South Asia—and redirection of attention and resources toward East Asia. These moves were responsive to the public mood. Most of the president's individual decisions pursuant to this strategy have had broad support, including the withdrawal of American forces from Iraq, starting to do the same in Afghanistan, "leading from behind" in Libya, and not intervening militarily in Syria. However, while each individual decision may have been popular, the overall results have not been. The American people may have been happy to assume less cost and fewer risks abroad but have not been satisfied with the resulting decrease in influence and increase in threatening disorder.

Overlearning the Lessons of the Recent Past

If one were to caricature the foreign policy of George W. Bush's first term as action without reflection, one might contrast Obama's approach as reflection without action. Neither charge is entirely just, but, certainly, Bush's greatest failing was one of commission (invading Iraq), while Obama's have been ones of omission (abandoning Iraq, not stabilizing Libya, and not doing anything about Syria before it became overrun by militant extremists). Obama's policy choices are the result of more than a decade's disappointment with counterinsurgency, nation-building, and democracy promotion.

It took the United States a generation to get over the lost war in Vietnam; it can ill afford to wait another generation before recovering from the setbacks in Iraq and Afghanistan and the disappointing results of the Arab Spring. This is not to say that America's leaders should not reflect on and absorb the lessons of the past decade, but those lessons are not that nation-building never works, that counterinsurgency is always too expensive, or that democracy promotion is ineffectual and potentially counterproductive because it sometimes produces worse governments than those it displaces.

Tomes have been written about the mistakes made in Afghanistan and Iraq during the Bush administration's first term. The United States went into both countries determined not to engage in nation-building, so it should be no surprise that its record in this regard disappoints. In both cases, the administration grossly underestimated the resources needed to stabilize these societies, eventually deploying the needed assets only years later, by which time large, well-organized resistance movements had emerged.

The lesson to be drawn from Afghanistan and Iraq is not that nation-building does not work but rather that it is expensive and time consuming. More than a dozen countries around the world today are at peace because U.S. troops—or NATO, European, United Nations, or African Union forces—intervened to end a civil war, provided security to the population, oversaw elections, helped install new governments, and then stayed around long enough to make sure the new regime took hold.[1] Few of these societies are prosperous, well governed, or fully democratic, but they are more prosperous, better governed, and more democratic. Most importantly, they are at peace with themselves and their neighbors, which was the prime objective of the interventions in the first place.

---

[1] Successful stabilization missions include Bosnia, Kosovo, Sierra Leone, East Timor, Mozambique, Liberia, El Salvador, Namibia, Cambodia, Croatia, Albania, Macedonia, Ivory Coast, and the Solomon Islands. See James Dobbins, Laurel E. Miller, Stephanie Pezard, Christopher S. Chivvis, Julie E. Taylor, Keith Crane, Calin Trenkov-Wermuth, and Tewodaj Mengistu, *Overcoming Obstacles to Peace: Local Factors in Nation-Building*, Santa Monica, Calif.: RAND Corporation, RR-167-CC, 2013.

A second lesson already evident in Iraq and Libya is that forced regime change that is not followed by successful stabilization operations can actually create a situation worse than the one the original intervention was intended to correct.

The objective of a stability operation is to prevent the renewal of conflict. Insurgency is what one gets if the stability operation fails, as happened almost immediately in Iraq and after several years in Afghanistan, or is never attempted, as in Libya. At that point, one must either counter the insurgency or allow it to prevail. Obviously, it is better to leave this to indigenous forces—assuming they exist—but the United States will not be able to help other regimes carry out successful counterinsurgency operations if it lacks the expertise and capacity to do so itself.

Democratization is not a binary condition. Some societies moved rapidly from authoritarian to democratic rule. In others, that process was much more gradual. Since the end of World War II, and particularly since the end of the Cold War, dozens of countries have moved away from the authoritarian camp, and democracy is now the dominant form of government throughout the world. There has been some limited regression over the past decade, with several Arab societies having transitioned from authoritarian government to none at all in the wake of the Arab Spring. Rather than giving up on the process, however, Americans need to temper their expectations and work to promote the foundations of democratic government—civil society, rule of law, growth of the middle class—so that when the next upheavals occur, as they inevitably will, the results will be more positive. We also need to recognize that almost any government is better than none.

U.S. National Strategy

Since becoming the world's most powerful nation in the middle of the last century, the United States has labored to build a rules-based global system dominated by market democracies and dedicated to promoting free trade and the peaceful settlement of disputes. This new order emerged from the Truman Doctrine, the Marshall Plan, and the reconstruction of Germany and Japan. It was bolstered by the creation of the United Nations, NATO, the World Bank, the IMF, and the WTO. It eventually produced the end of colonialism, the reunification of Germany, the liberation of Eastern Europe, the peaceful dissolution of the Soviet empire, the extension of democracy to 144 countries,[2] and the greatest advances in global longevity and poverty reduction in human experience. Preserving and extending this world order should remain the core objective of American policy—first, because it provides the best environment for the United States' own security and prosperity and, second, because if the United States does not lead, no one else will. Defining a national strategy toward this end requires the identification of ends, ways, and means.

Ends

American interests and values will be best served in a world in which states adhere to established norms of behavior that ensure peace and promote prosperity. Importantly, these same states must remain capable and willing to ensure that their citizens observe such norms. Threats to this order are thus posed by both states that flout the rules and states that prove incapable of enforcing them. Any unwillingness or incapacity to exercise the responsibilities of sovereignty becomes all the more dangerous in a world ever more closely knit by trade, travel, and communications, a world in which mere physical proximity is no longer the prime factor in determining the source of serious threats.

Today's state-based threats to international order come primarily from Russia; Iran; North Korea; and, potentially, China. Russia's aggression against Ukraine needs to be made costly enough to discourage repetition or imitation, and barriers against further Russian aggression should be strengthened. China needs to be deterred from any similar behavior. We must do what we can to prevent Iran from acquiring nuclear weapons. North Korea must be prevented from proliferating its nuclear capacity and dissuaded from attacking its neighbors. China's

[2] Freedom House finds 88 countries free and 59 partly free, representing a decline over the past decade, but still much higher than in any other era. See Freedom House, *Freedom in the World 2014*, Washington, D.C., 2014.

growing power needs to be channeled constructively. One can pose and even pursue more-ambitious goals, such as turning back Russian aggression, fully dismantling Iran's nuclear infrastructure, denuclearizing Korea, and preventing an expansion of Chinese influence, but the costs and risk associated with these more-ambitious aims may be prohibitive.

Today's nonstate threats come principally from violent extremist groups based in the Middle East, North Africa, and South Asia. Eliminating these groups entirely is beyond U.S. capacity, but they should at least be suppressed to an extent that they no longer hold territory and govern large populations, are no longer able to do great damage at great distance, and no longer recruit and inspire large numbers of individuals from within Western and other societies. Ending the civil war in Syria, even at the expense of dealing with Assad, is probably the single most important step toward suppressing the most virulent of these groups and diminishing their ability to attract new recruits and inspire imitators. This will require engaging Iran while also maintaining existing U.S. partnerships and alliances.

The defensive aspects of American strategy must be to punish past and deter future state-on-state aggression while suppressing violent nonstate extremist movements that threaten U.S. citizens and those of our friends and allies. For these efforts to have any lasting effect, the United States also needs to sustain and extend a rules-based international order founded on states that are willing to observe certain norms of behavior and able to ensure that their citizens do likewise. A major goal of U.S. policy should therefore be to improve the capacity of international institutions to channel collective action; to strengthen the capacity of individual states to engage effectively in such action; and to further develop international norms in new areas of vulnerability, most notably in the cyber and climate domains, where an adequate body of generally accepted rules are currently lacking.

Ways

The ways of dealing with threats from strong states are conceptually well developed, if expensive and demanding in application. These include diplomacy, economic pressure, deterrence, containment, and collective defense.

The ways of dealing with threats emanating from weak states include CT and counterinsurgency operations combined with state- and nation-building efforts.

Counterterrorism and counterinsurgency campaigns can disrupt extremist groups, but only the establishment of states willing to adhere to and capable of enforcing international norms will allow such campaigns to be concluded. State- and nation-building operations are long-term, resource-intensive enterprises, but at least they have a defined end state. Counterterror operations do not.

In today's hyperlinked world, few challenges are amenable to unilateral responses. American involvement will be essential to the resolution of the many challenges outlined in this book, but it is seldom sufficient. Slowing global climate change is only the most extreme example. Deterring or turning back aggression almost always requires collective defense. State- and nation-building efforts require support from neighbors and near neighbors, from the societies that have the most at stake and possess the most influence by reason of their proximity and commercial, criminal, religious, and cultural connections. Partnerships are essential. Coalitions are the norm. One cannot afford to be too choosy about the company one keeps. Russia is needed to help prevent Iran from acquiring nuclear weapons. China is needed to restrain North Korea. Iran is needed to fight ISIS. Both Russia and Iran will be needed to end the war in Syria. Stemming climate change will require almost global efforts. Partnerships in this era are not just about friends confronting enemies.

More government may not be the answer to America's domestic problems, but more and better government abroad is certainly essential to meet most of its external challenges. State-building will therefore remain an unavoidable mission for the United States in Europe, the Middle East, and South Asia, as well as in Africa. This need not always involve military interventions, nor must the United States be the principal source of military manpower and economic assistance. Where regional states have been willing to take the lead, as in Africa, U.S. advice and financial assistance may be all that is necessary. Building a functional Ukrainian state and continuing to stabilize the Balkans will be tasks primarily for Europe. But the United States should not base its defense planning and size its own military establishment on the premise that it is permanently out of the nation-building business. There is nothing in our history or our prospective future that supports such an assumption.

Means

As a proportion of national income, the U.S. defense budget is headed for its lowest level in some 50 years. Yet with the continued rise of China, the emergence of ISIS, and intensified Russian belligerence, the United States now faces the need to fight, or at least deploy forces sufficient to fight, on three fronts against at least three different opponents. The current budget and envisaged force structure are inadequate to this task, as successive U.S. Secretaries of Defense have acknowledged.

During the Cold War, the United States sized and organized its military establishment to be able to fight and win two major wars at once, one in Asia and one in Europe, against two different opponents. In recent decades, this requirement has been nominally sustained, but the scale of each envisaged conflict has been reduced, reflecting both a diminished threat and reduced capacity. The current standard is to defeat one regional adversary in a large-scale, multiphased campaign and deny the objective of, or impose unacceptable costs on, a different aggressor in another region.[3] Additionally, discouraged by the results of its efforts in Iraq and Afghanistan, the Obama administration has decided not to require U.S. armed forces to be ready to conduct large-scale, protracted stability or counterinsurgency operations and has cut the size of the Army and Marine Corps accordingly. Meanwhile, many modernization programs have been slowed or truncated, and readiness has eroded due to reduced levels of defense funding.

These trends are clearly incompatible with the three-theater challenge the United States now faces. The current international environment does not permit the United States to continue to transfer its time, attention, and national resources from West to East. Further military reductions from Europe, the Middle East, and South Asia to East Asia

---

[3] DoD, 2012.

are not prudent. The current decline in the U.S. defense budget—and the national security budget more generally—must be slowed, and all three theaters need to be adequately resourced.

Within American society, the predominant but never-uncontested tendency for most of the period since World War II has favored American leadership in the sustainment and expansion of a rules-based international order promoting free trade and the peaceful settlement of disputes, combined with a willingness to commit diplomatic, economic, and military resources to develop and help enforce such norms.

A more-hawkish camp has been equally committed to American global leadership and even more ready to devote resources to that effort but places somewhat less emphasis on multilateralism and the further elaboration of enforceable norms of behavior.

A third group feels the United States is overextended abroad, does not believe the United States needs to assume the dominant responsibility for addressing every challenge to global order, and would like to reduce the overall level of U.S. overseas commitments. On the political right, this group includes small-government conservatives and libertarians. On the left, its supporters include those who prefer "nation-building at home," generally oppose the use of force, and resist further globalization.

There remains, nevertheless, a sizable constituency in Congress, as in the public, in favor of a well-resourced foreign and security policy agenda, but it spans both parties and dominates neither. This constituency can prevail only if there is willingness on both sides to cross party lines and support sensible policies. As long as the two parties remain dug in behind the partisan ramparts of "no new taxes" and "no cuts to entitlements," the United States' external problems will mount as new challenges pile on top of old ones that have not been addressed. As a result, the global order will indeed begin to erode, and fundamental American interests will suffer.

Conclusion

The United States, as the world's most powerful nation, should continue to take the lead in sustaining and extending a rules-based international order. It should promote the development of new norms in domains where these do not yet exist, such as cyber and climate management. States are the essential building blocks in any such system. Challenges come from strong states that break the rules and from weak ones that cannot enforce them. Both these challenges need to be addressed. A focus on defense, deterrence, and dissuasion is essential, but it is not enough. State capacity needs to keep pace with the growing capacity for disruption of individuals and groups. The most successful eras of American statecraft have been periods of construction: the birth of new institutions, the reconstruction of shattered nations, and the establishment of new norms for international behavior. The United States needs to combine its defense of existing institutions and norms with a rededication to such a positive agenda and to commit itself to providing the necessary resources.

Abbreviations

A2/AD	antiaccess and area denial
AIIB	Asian Infrastructure Investment Bank
BCA	Budget Control Act
BIT	bilateral investment treaty
BRICS	Brazil, Russia, India, China, and South Africa
CBO	Congressional Budget Office
CT	counterterrorism
DoD	Department of Defense
EU	European Union
FDI	foreign direct investment
FTA	free trade agreement
FY	fiscal year
GDP	gross domestic product
GtC	gigatons of carbon
ICBM	intercontinental ballistic missile
ICT	information and communications technology
IMF	International Monetary Fund

IoT	Internet of Things
ISIS	Islamic State of Iraq and Syria
LRSO	Long Range Standoff Weapon
NARA	nuclear-armed regional adversary
NATO	North Atlantic Treaty Organization
NSA	National Security Agency
SOF	special operations forces
SSBN	nuclear-powered ballistic missile submarine
TPP	Trans-Pacific Partnership
TTIP	Trans-Atlantic Trade and Investment Partnership
UNFCCC	United Nations Framework Convention on Climate Change
WTO	World Trade Organization

References

Akhtar, Shayerah Ilias, and Vivian C. Jones, *Proposed Transatlantic Trade and Investment Partnership (T-TIP): In Brief*, Washington, D.C.: Congressional Research Service, June 11, 2014.

Allen, Myles R., David J. Frame, Chris Huntingford, Chris D. Jones, Jason A. Lowe, Malte Meinhausen, and Nicolai Meinshausen, "Warming Caused by Cumulative Carbon Emissions Towards the Trillionth Tonne," *Nature*, Vol. 458, 2009, pp. 1163–1166.

Asia-Pacific Partnership on Clean Development and Climate, homepage, undated. As of June 1, 2015:
http://www.asiapacificpartnership.org/english/default.aspx

Aylett, Alexander, *Progress and Challenges in the Urban Governance of Climate Change: Results of a Global Survey*, Cambridge, Mass.: Massachusetts Institute of Technology, 2014.

Bamford, James, and Tim De Chant, "Exclusive: Edward Snowden on Cyber Warfare," interview, NOVA, PBS, January 8, 2015. As of May 5, 2015:
http://www.pbs.org/wgbh/nova/next/military/snowden-transcript/

Belasco, Amy, *Defense Spending and the Budget Control Act Limits*, Washington, D.C.: Congressional Research Service, R44039, May 19, 2015.

Bloomberg, M. R., H. M. Paulson, and T. F. Steyer, *Risky Business: The Economic Risks of Climate Change in the United States*, Risky Business Project, 2014.

Branigan, Tania, "South Korea on Alert for Cyber-Attacks After Major Network Goes Down," *The Guardian*, March 20, 2013. As of February 19, 2015:
http://www.theguardian.com/world/2013/mar/20/south-korea-under-cyber-attack

Bureau of Economic Analysis, "Percent Change from Preceding Period," spreadsheet, U.S. Department of Commerce, May 29, 2015. As of June 14, 2015:
http://www.bea.gov/national/index.htm#gdp

Burroughs, W. J., *Climate Change in Prehistory: The End of the Reign of Chaos*, Cambridge, UK: Cambridge University Press, 2005.

Bush, George H. W., "Remarks to the Citizens in Mainz," May 31, 1989.

California Environmental Protection Agency, Air Resources Board, "California Greenhouse Gas Emission Inventory," web page, updated December 22, 2014. As of June 1, 2015:
http://www.arb.ca.gov/cc/inventory/inventory_current.htm

Carter, Ash, "Remarks by Secretary Carter at the Drell Lecture Cemex Auditorium, Stanford Graduate School of Business," transcript, Stanford, Calif., April 23, 2015. As of May 1, 2015:
http://www.defense.gov/Transcripts/Transcript.aspx?TranscriptID=5621

CBO—*See* Congressional Budget Office.

Center for Strategic and International Studies, *Net Losses: Estimating the Global Cost of Cybercrime*, Washington, D.C., June 2014. As of May 27, 2015:
http://csis.org/files/attachments/140609_rp_economic_impact_cybercrime_report.pdf

Chase, Michael S., "Chinese Suspicion and U.S. Intentions," *Survival*, June–July 2011, pp. 133–150.

Chrisafis, Angelique, "France 'Runs Vast Electronic Spying Operation Using NSA-Style Methods,'" *The Guardian*, July 4, 2013. As of May 27, 2015:
http://www.theguardian.com/world/2013/jul/04/france-electronic-spying-operation-nsa

Cisco, "Seize New Product and Revenue Opportunities with the Internet of Things," web page, undated. As of May 1, 2015:
http://www.cisco.com/web/solutions/trends/iot/portfolio.html

Cliff, Roger, Mark Burles, Michael S. Chase, Derek Keaton, and Kevin L. Pollpeter, *Entering the Dragon's Lair: Chinese Antiaccess Strategies and Their Implications for the United States*, Santa Monica, Calif.: RAND Corporation, MG-524-AF, 2007. As of May 26, 2015:
http://www.rand.org/pubs/monographs/MG524.html

Cliff, Roger, Phillip C. Saunders, and Scott Warren Harold, eds., *New Opportunities and Challenges for Taiwan's Security*, Santa Monica, Calif.: RAND Corporation, CF-279-OSD, 2011. As of May 27, 2015:
http://www.rand.org/pubs/conf_proceedings/CF279.html

Colson, Mia, Kristin Heery, and Allan Wallis, *A Survey of Regional Planning for Climate Adaptation*, Washington, D.C.: National Association of Regional Councils, undated.

Congressional Budget Office, *Projected Costs of U.S. Nuclear Forces, 2014 to 2023*, Washington, D.C., December 2013. As of May 27, 2015:
http://www.cbo.gov/sites/default/files/cbofiles/attachments/12-19-2013-NuclearForces.pdf

———, *The 2014 Long-Term Budget Outlook*, Washington, D.C., July 2014a.

———, *An Update to the Budget and Economic Outlook: 2014 to 2024*, Washington, D.C., August 2014b.

Cooper, Richard, "The Third Plenum and Economic Reform," *Harvard Economics Review*, Spring 2014, pp. 12–14.

Council of Europe, Convention on Cybercrime, Budapest, November 23, 2001. As of March 14, 2015:
http://conventions.coe.int/Treaty/en/Treaties/Html/185.htm

CrowdStrike, *Putter Panda: PLA 3rd Department 12th Bureau Unit 61486*, June 9, 2014. As of June 12, 2014:
http://resources.crowdstrike.com/putterpanda/

"The Decision on Major Issues Concerning Comprehensively Deepening Reforms in Brief," *China Daily*, November 16, 2013.

Defense Science Board Task Force, *Trends and Implications for Climate Change on National and International Security*, Washington, D.C.: U.S. Department of Defense, Office of the Under Secretary of Defense for Acquisition, Technology, and Logistics, October 2011.

Department of Defense, *Sustaining U.S. Global Leadership: Priorities for 21st Century Defense*, Washington, D.C., January 2012. As of May 19, 2015:
http://www.defense.gov/news/Defense_Strategic_Guidance.pdf

———, *The DoD Cyber Strategy*, Washington, D.C., April 2015. As of May 11, 2015:
http://www.defense.gov/home/features/2015/0415_cyber-strategy/Final_2015_DoD_CYBER_STRATEGY_for_web.pdf

Department of Homeland Security, "What Is Critical Infrastructure?" web page, October 24, 2013. As of February 17, 2015:
http://www.dhs.gov/what-critical-infrastructure

Dobbins, James, Laurel E. Miller, Stephanie Pezard, Christopher S. Chivvis, Julie E. Taylor, Keith Crane, Calin Trenkov-Wermuth, and Tewodaj Mengistu, *Overcoming Obstacles to Peace: Local Factors in Nation-Building*, Santa Monica, Calif.: RAND Corporation, RR-167-CC, 2013. As of June 30, 2015:
http://www.rand.org/pubs/research_reports/RR167.html

DoD—*See* Department of Defense.

Dollar, David, "China's Rebalancing: Lessons from East Asian Economic History," Working Paper Series, Washington, D.C.: John L. Thornton China Center, Brookings Institution, October 2013.

Dozier, Kimberly, "Former DIA Chief Flynn Calls for Global War on Islamic Extremists," *Daily Beast*, February 13, 2015. As of June 14, 2015:
http://www.thedailybeast.com/articles/2015/02/13/former-dia-chief-flynn-calls-for-global-war-on-islamic-extremists.html

Duchatel, Mathieu, "Tactical Pause in China's Economic Engagement with North Korea," *China Brief*, Vol. 14, No. 9, May 7, 2014.

European Commission, "Countries and Regions: Japan," web page, Directorate-General for Trade, September 15, 2014. As of October 8, 2014:
http://ec.europa.eu/trade/policy/countries-and-regions/countries/japan/

Ewing, Jack, and Gaia Pianigiana, "Italy Falls Back into Recession, Raising Concern for Eurozone Economy," *New York Times*, August 6, 2014. As of June 1, 2015:
http://www.nytimes.com/2014/08/07/business/international/italy-falls-back-into-recession-raising-concern-for-eurozone-economy.html

Executive Order 12333, United States Intelligence Activities, December 4, 1981. As of February 18, 2015:
http://www.archives.gov/federal-register/codification/executive-order/12333.html

Fergusson, Ian F., William H. Cooper, Remy Jurenas, and Brock R. Williams, *The Trans-Pacific Partnership (TPP) Negotiations and Issues for Congress*, Washington, D.C.: Congressional Research Service, December 13, 2013.

Freedman, Andrew, "The Last Time CO2 Was This High, Humans Didn't Exist," *Climate Central*, May 3, 2013. As of June 17, 2015:
http://www.climatecentral.org/news/
the-last-time-co2-was-this-high-humans-didnt-exist-15938

Freedom House, *Freedom in the World 2014*, Washington, D.C., 2014. As of July 15, 2015:
https://freedomhouse.org/report/freedom-world/freedom-world-2014

Friedberg, Aaron, "Bucking Beijing: An Alternative U.S. China Policy," *Foreign Affairs*, September/October 2012.

Giles, Martin, "Defending the Digital Frontier," *The Economist*, July 12, 2014.

Gillis, Justin, "A Tricky Transition from Fossil Fuel: Denmark Aims for 100 Percent Renewable Energy," *New York Times*, November 14, 2014.

Gompert, David C., and Terrence K. Kelly, "Escalation Clause: How the Pentagon's New Strategy Could Trigger War with China," *Foreign Policy*, August 2, 2013.

Goodin, Dan, "How 'Omnipotent' Hackers Tied to NSA Hid for 14 Years—and Were Found at Last," Ars Technica, February 16, 2015. As of February 17, 2015:
http://arstechnica.com/security/2015/02/
how-omnipotent-hackers-tied-to-the-nsa-hid-for-14-years-and-were-found-at-last/

Goossens, Ehren, and M. Chediak, "Battery-Stored Solar Power Sparks Backlash from Utilities," *Bloomberg*, October 8, 2013. As of June 1, 2015:
http://www.bloomberg.com/news/articles/2013-10-07/
battery-stored-solar-power-sparks-backlash-from-utilities

Green, Michael J., Gregory Kiley, Nicholas Szechenyi, and David J. Berteau, *U.S. Force Posture Strategy in the Asia Pacific Region: An Independent Assessment*, Washington, D.C.: Center for Strategic and International Studies, 2013.

Hagel, Chuck, "IISS Shangri-La Dialogue," speech, Singapore, May 31, 2014. As of June 1, 2015:
http://www.defense.gov/Speeches/Speech.aspx?SpeechID=1857

Harvey, Hal, Franklin M. Orr, Jr., and Clara Vondrich, "A Trillion Tons," *Daedalus*, Vol. 142, No. 1, Winter 2013, pp. 8–25.

Hruska, Joel, "The NSA Wants 'Front Door' Access to Your Encrypted Data," *ExtremeTech*, April 13, 2015. As of May 12, 2015:
http://www.extremetech.com/extreme/203275-the-nsa-wants-front-door-access-to-your-encrypted-data

IMF—*See* International Monetary Fund.

Intergovernmental Panel on Climate Change, *Climate Change 2013: The Physical Science Basis*, Cambridge, UK: Cambridge University Press, 2013.

———, *Climate Change 2014: Impacts, Adaptation, and Vulnerability*, Cambridge, UK: Cambridge University Press, 2014a.

———, *Climate Change 2014: Mitigation of Climate Change*, Cambridge, UK: Cambridge University Press, 2014b.

International Institute for Strategic Studies, "North America," *The Military Balance*, London: Routledge, 2014, pp. 31–58.

International Monetary Fund, *World Economic Outlook: Legacies, Clouds, Uncertainties*, World Economic and Financial Surveys, Washington, D.C., October 2014.

———, *World Economic Outlook Update*, Washington, D.C., January 2015a.

———, "IMF and China's Ministry of Finance Sign Agreement on Strengthening Fiscal Institutions and Capacity Development," press release, Washington, D.C., No. 15/141, March 27, 2015b. As of April 7, 2015:
http://www.imf.org/external/np/sec/pr/2015/pr15141.htm

Jones, Seth G., *A Persistent Threat: The Evolution of al Qa'ida and Other Salafi Jihadists*, Santa Monica, Calif.: RAND Corporation, RR-637-OSD, 2014. As of May 26, 2015:
http://www.rand.org/pubs/research_reports/RR637.html

Kennan, George F., *Memoirs 1925–1950*, New York: Pantheon, 1983.

Kerry, John, "Remarks on U.S. China Relations," Johns Hopkins School of Advanced International Studies, Washington, D.C., November 4, 2014. As of June 1, 2015:
http://www.state.gov/secretary/remarks/2014/11/233705.htm

Kissinger, Henry A., *Global Challenges and U.S. National Security Strategy*, testimony before the U.S. Senate Armed Services Committee, Washington, D.C., January 20, 2015.

Landler, Mark, "Latest North Korean Mystery: A Diplomatic Charm Offensive," *New York Times*, October 23, 2014a.

———, "US and China Reach Climate Accord After Months of Talks," *New York Times*, November 11, 2014b. As of June 1, 2015:
http://www.nytimes.com/2014/11/12/world/asia/china-us-xi-obama-apec.html?_r=0

Leclère, D., P. Havlík, S. Fuss, A. Mosnier, E. Schmid, H. Valin, M. Herrero, N. Khabarov, and M. Obersteine, "Climate Change Impacts on Agriculture, Adaptation and the Role of Uncertainty," *Impacts World 2013*, Potsdam: International Conference on Climate Change Effects, May 27–30, 2013.

Liberthal, Kenneth, and Wang Jisi, *Addressing U.S.-China Strategic Distrust*, Washington, D.C.: Brookings Institution, March 2012. As of June 1, 2015:
http://www.brookings.edu/~/media/Research/Files/Papers/2012/3/30%20us%20china%20lieberthal/0330_china_lieberthal.pdf

Libicki, Martin, *Cyberdeterrence and Cyberwar*, Santa Monica, Calif.: RAND Corporation, MG-877-AF, 2009. As of May 27, 2015:
http://www.rand.org/pubs/monographs/MG877.html

Maddison, Angus, *The World Economy: A Millennial Perspective*, Paris: Organisation for Economic Co-operation and Development, 2001.

Madeiros, Evan S., Keith Crane, Eric Heginbotham, Norman D. Levin, Julia F. Lowell, Angel Rabasa, and Somi Seong, *Pacific Currents: The Responses of U.S. Allies and Security Partners in East Asia to China's Rise*, Santa Monica, Calif.: RAND Corporation, MG-736-AF, 2008. As of June 1, 2015:
http://www.rand.org/pubs/monographs/MG736.html

Major Economies Forum on Energy and Climate, homepage, 2009. As of June 1, 2015:
http://www.majoreconomiesforum.org

Mandiant Intelligence Center, *APT1: Exposing One of China's Cyber Espionage Units*, undated. As of February 15, 2015:
http://intelreport.mandiant.com/Mandiant_APT1_Report.pdf

Markoff, John, "Before the Gunfire, Cyberattacks," *New York Times*, August 12, 2008. As of February 18, 2015:
http://www.nytimes.com/2008/08/13/technology/13cyber.html?_r=0

McKinsey and Company, *Reducing U.S. Greenhouse Gas Emissions: How Much at What Cost?* U.S. Greenhouse Gas Abatement Mapping Initiative, Executive Report, December 2007.

———, *Impact of the Financial Crisis on Carbon Economics: Version 2.1 of the Global Greenhouse Gas Abatement Cost Curve*, 2010.

Mercier, Keith, "The Internet of Things Will Transform Retail as We Know It," *Forbes*, January 12, 2015. As of May 1, 2015:
http://www.forbes.com/sites/ibm/2015/01/12/the-internet-of-things-will-transform-retail-as-we-know-it/

Nakashima, Ellen, "Russian Hackers Use 'Zero-Day' to Hack NATO, Ukraine in Cyber-Spy Campaign," *Washington Post*, October 13, 2014. As of February 18, 2015:
http://www.washingtonpost.com/world/national-security/russian-hackers-use-zero-day-to-hack-nato-ukraine-in-cyber-spy-campaign/2014/10/13/f2452976-52f9-11e4-892e-602188e70e9c_story.html

Nakashima, Ellen, and Barton Gellman, "As Encryption Spreads, U.S. Grapples with Clash Between Privacy, Security," *Washington Post*, April 10, 2015. As of May 12, 2015:
http://www.washingtonpost.com/world/national-security/as-encryption-spreads-us-worries-about-access-to-data-for-investigations/2015/04/10/7c1c7518-d401-11e4-a62f-ee745911a4ff_story.html

Nathan, Andrew J., and Andrew Scobell, "How China Sees America: The Sum of Beijing's Fears," *Foreign Affairs*, September/October 2012.

National Institute of Standards and Technology, *Framework for Improving Critical Infrastructure Cybersecurity*, Version 1.0, February 12, 2014. As of May 12, 2015:
http://www.nist.gov/cyberframework/upload/cybersecurity-framework-021214.pdf

National Science Board, *Science and Engineering Indicators 2014*, Arlington, Va.: National Science Foundation, NSB 14-01, 2014.

NATO—*See* North Atlantic Treaty Organization.

NATO Cooperative Cyber Defence Centre of Excellence, *Tallinn Manual on the International Law Applicable to Cyber Warfare*, Cambridge, UK: Cambridge University Press, 2013.

Nordhaus, William D., "A New Solution: The Climate Club," *The New York Review of Books*, June 4, 2015. As of June 23, 2015:
http://www.nybooks.com/articles/archives/2015/jun/04/new-solution-climate-club/

North Atlantic Treaty Organization, Public Diplomacy Division, "Financial and Economic Data Relating to NATO Defence," press release 28, February 24, 2014. As of May 27, 2015:
http://www.nato.int/nato_static/assets/pdf/pdf_topics/20140224_140224-PR2014-028-Defence-exp.pdf

North Korea Inside Out: The Case for Economic Engagement, Report of an Independent Task Force convened by the Asia Society's Center on U.S.-China Relations and the University of California Institute on Global Conflict and Cooperation, December 2009. As of June 1, 2015:
http://asiasociety.org/files/pdf/North_Korea_Inside_Out.pdf

Obama, Barack, "Remarks by the President on Review of Signals Intelligence," Washington, D.C.: The White House, Office of the Press Secretary, January 17, 2014a. As of June 30, 2015:
https://www.whitehouse.gov/the-press-office/2014/01/17/remarks-president-review-signals-intelligence

———, "Remarks by President Obama to the People of Estonia," Washington, D.C.: The White House, Office of the Press Secretary, September 3, 2014b. As of June 30, 2015:
https://www.whitehouse.gov/the-press-office/2014/09/03/remarks-president-obama-people-estonia

———, "Remarks by the President at the Federal Trade Commission," Washington, D.C.: The White House, Office of the Press Secretary, January 12, 2015. As of May 4, 2015:
https://www.whitehouse.gov/the-press-office/2015/01/12/remarks-president-federal-trade-commission

Ochmanek, David, and Lowell H. Schwartz, *The Challenge of Nuclear-Armed Regional Adversaries*, Santa Monica, Calif.: RAND Corporation, MG-671-AF, 2008. As of May 27, 2015:
http://www.rand.org/pubs/monographs/MG671.html

Office of Management and Budget, "Historical Tables," Washington, D.C.: The White House, undated. As of May 27, 2015:
www.whitehouse.gov/omb/budget/Historicals

Office of the Secretary of Defense, *Military and Security Developments Involving the Democratic People's Republic of Korea*, Washington, D.C.: U.S. Department of Defense, 2014. As of June 1, 2015:
http://www.defense.gov/pubs/North_Korea_Military_Power_Report_2013-2014.pdf

Office of the Under Secretary of Defense (Comptroller)/Chief Financial Officer, *United States Department of Defense Fiscal Year 2015 Budget Request Overview*, Washington, D.C.: U.S. Department of Defense, March 2014.

Orcutt, Mike, "The United States Is Far and Away the Leader in Carbon Dioxide Emissions," *MIT Technology Review*, November 12, 2014.

"Pacific Coast Action Plan on Climate and Energy," San Francisco, Calif., signed October 28, 2013. As of June 1, 2015:
http://www.pacificcoastcollaborative.org/Documents/Pacific%20Coast%20Climate%20Action%20Plan.pdf

Pacific Coast Collaborative, homepage, undated. As of June 1, 2015:
http://www.pacificcoastcollaborative.org/Pages/Welcome.aspx

Parry, Ian W. H., and Roberton C. Williams III, *Moving U.S. Climate Policy Forward: Are Carbon Taxes the Only Good Alternative?* Washington, D.C.: Resources for the Future, Discussion Paper RFF DP 11-02, February 2011.

Perry, William J., and John P. Abizaid, *Ensuring a Strong U.S. Defense for the Future: The National Defense Panel Review of the 2014 Quadrennial Defense Review*, Washington, D.C.: United States Institute of Peace, July 2014. As of May 27, 2015:
http://www.usip.org/sites/default/files/Ensuring-a-Strong-U.S.-Defense-for-the-Future-NDP-Review-of-the-QDR.pdf

Peterson, Larry, and Gerald Haug, "Climate Change and the Collapse of Mayan Civilization," *American Scientist*, Vol. 93, 2005, pp. 322–329.

Pew Research Center, "Growing Support for Campaign Against ISIS—and Possible Use of U.S. Ground Troops," February 24, 2015. As of June 14, 2015:
http://www.people-press.org/2015/02/24/
growing-support-for-campaign-against-isis-and-possible-use-of-u-s-ground-troops/

Pollack, Jonathan, *No Exit: North Korea, Nuclear Weapons, and International Security*, London: International Institute for Strategic Studies, May 2011.

Ratner, Ely, "Rebalancing to Asia with an Insecure China," *Washington Quarterly*, Spring 2013, pp. 21–38.

Reader, Ruth, "Amazon's Drones Could Track You Down to Deliver Your Goods," *VB News*, May 8, 2015. As of May 11, 2015:
http://venturebeat.com/2015/05/08/
amazons-drones-could-track-you-down-to-deliver-your-goods/

Richards, Jason, "Denial-of-Service: The Estonian Cyberwar and Its Implications for U.S. National Security," *International Affairs Review*, Vol. 18, No. 2, 2009. As of May 28, 2015:
http://www.iar-gwu.org/node/65

Royal United Services Institute, *Cyber Norms of Behaviour: Executive Summary*, London, undated. As of March 15, 2015:
https://www.rusi.org/downloads/assets/Cyber_norms_of_behaviour_report_-_Executive_Summary.pdf

Rubin, Alissa J., and David E. Sanger, "Familiar Swing to Security Over Privacy After Attacks in France," *New York Times*, May 6, 2015. As of May 6, 2015:
http://www.nytimes.com/2015/05/07/world/europe/france-expanded-surveillance-charlie-hebdo.html?_r=1

Sanger, David E., and Martin Fackler, "N.S.A. Breached North Korean Networks Before Sony Attack, Officials Say," *New York Times*, January 18, 2015. As of January 25, 2015:
http://www.nytimes.com/2015/01/19/world/asia/nsa-tapped-into-north-korean-networks-before-sony-attack-officials-say.html

Schmitt, Eric, and Steven Lee Myers, "U.S. Is Poised to Put Heavy Weaponry in Eastern Europe," *New York Times*, June 13, 2015. As of June 15, 2015:
http://www.nytimes.com/2015/06/14/world/europe/us-poised-to-put-heavy-weaponry-in-east-europe.html?_r=0

Schogol, Jeff, "Less Than Half of Combat Squadrons Fully Ready for Combat," *Air Force Times*, March 4, 2015. As of May:
http://www.airforcetimes.com/story/military/2015/02/25/fewer-than-half-of-combat-squadrons-fully-ready-for-combat/23997409/

Schott, Jeffrey J., "Can Mega-Regionals Support Multilateralism?" presented at the OECD Global Forum on Trade, Reconciling Regionalism and Multilateralism in a Post-Bali World, Paris, February 11, 2014.

Scobell, Andrew, and Mark Cozad, "China's North Korea Policy: Rethink or Recharge?" *Parameters*, Vol. 44, No. 1, Spring 2014, pp. 51–63.

Smil, Vaclav, *Energy Transitions: History, Requirements, Prospects*, Santa Barbara, Calif.: Praeger, 2010.

Stokes, Mark A., Jenny Lin, and L. C. Russell Hsiao, *The Chinese People's Liberation Army Signals Intelligence and Cyber Reconnaissance Infrastructure*, Project 2049 Institute, November 11, 2011. As of June 12, 2014:
http://project2049.net/documents/pla_third_department_sigint_cyber_stokes_lin_hsiao.pdf

Subramanian, Arvind, "Preserving the Open Global Economic System: A Strategic Blueprint for China and the United States," policy brief, Washington, D.C.: Peterson Institute for International Economics, PB13-16, June 2013.

Tattersall, Ian, and Jeffrey Schwartz, *Extinct Humans*, Boulder, Colo.: Westview Press, 2001.

"Towards the End of Poverty," *The Economist*, June 1, 2013.

United Nations Framework Convention on Climate Change, "Copenhagen Accord," FCCC/CP/2009/L.7, Draft Decision -/CP.15, December 18, 2009.

———, "Report of the Conference of the Parties on Its Fifteenth Session, Held in Copenhagen from 7 to 19 December 2009," Addendum, "Part Two: Action Taken by the Conference of the Parties at Its Fifteenth Session," FCCC/CP/2009/11/Add.1, March 30, 2010.

U.S. Agency for International Development, *USAID Climate Change and Development Strategy, 2012–2016: Clean Resilient Growth*, Washington, D.C., January 2012.

U.S. Department of Energy, *Quadrennial Energy Review: Energy Transmission, Storage and Distribution Infrastructure*, Washington, D.C., April 2015.

U.S. Environmental Protection Agency, *Inventory of U.S. Greenhouse Gas Emissions and Sinks: 1990–2013*, Washington, D.C., EPA 430-4-15-004, April 15, 2015.

U.S. Government Accountability Office, *Federal Information Security: Mixed Progress in Implementing Program; Improved Metrics Needed to Measure Effectiveness*, September 26, 2013. As of March 1, 2015:
http://www.gao.gov/products/GAO-13-776

———, *Climate Change Adaptation: DOD Can Improve Infrastructure Planning and Processes to Better Account for Potential Impacts*, Washington, D.C., GAO-14-446, May 2014.

Van den Hengel, Ruben, "The Rise of the Mega-FTAs," fact sheet, Singapore: EU Centre, October 2013.

Verizon, *2012 Data Breach Investigations Report*, March 2012. As of February 20, 2015:
http://www.verizonenterprise.com/resources/reports/rp_data-breach-investigations-report-2012-ebk_en_xg.pdf

Victor, D. G., J. House, and S. Joy, "A Madisonian Approach to Climate Policy," *Science*, Vol. 309, 2005, pp. 1820–1821.

Victor, David G., and Charles F. Kennel, "Climate Policy: Ditch the 2°C Warming Goal," *Nature*, Vol. 514, October 2, 2014, pp. 30–31.

Walker, Dinah, *Trends in U.S. Military Spending*, Washington, D.C.: Council on Foreign Relations, July 15, 2014. As of May 27, 2015:
http://www.cfr.org/defense-budgets/trends-us-military-spending/p28855

Weaver, C. P., R. J. Lempert, C. Brown, J. A. Hall, D. Revell, and D. Sarewitz, "Improving the Contribution of Climate Model Information to Decision Making: The Value and Demands of Robust Decision Frameworks," *Wiley Interdisciplinary Reviews: Climate Change*, Vol. 4, No. 1, 2013, pp. 39–60.

Weitz, Richard, "North Korea: The Problem with Reconciliation Via Engagement," *The Diplomat*, January 23, 2014. As of June 1, 2015:
http://thediplomat.com/2014/01/north-korea-the-problem-with-reconciliation-via-engagement/

Welsh, Mark A., "The Impact of Sequestration on National Defense," statement before the Senate Armed Services Committee, November 7, 2013. As of May 27, 2015:
http://www.armed-services.senate.gov/imo/media/doc/Welsh_11-07-13.pdf

The White House, Office of the Press Secretary, "Fact Sheet: U.S.-China Joint Announcement on Climate Change and Clean Energy Cooperation," Washington, D.C., November 11, 2014. As of June 1, 2015:
https://www.whitehouse.gov/the-press-office/2014/11/11/
fact-sheet-us-china-joint-announcement-climate-change-and-clean-energy-c

———, "Fact Sheet: U.S. Reports Its 2025 Emissions Target to the UNFCCC," Washington, D.C., March 31, 2015. As of June 1, 2015:
https://www.whitehouse.gov/the-press-office/2015/03/31/
fact-sheet-us-reports-its-2025-emissions-target-unfccc

The World Bank, "World Development Indicators," web page, April 14, 2015. As of June 14, 2015:
http://data.worldbank.org/data-catalog/world-development-indicators

World Resources Institute, "CAIT [Climate Analysis Indicators Tool] Climate Data Explorer," homepage, undated. As of June 1, 2015:
http://cait.wri.org

"Xinhua Insight: China Marks Six Priorities for New Type of Major-Country Relations with U.S.," *Xinhua*, November 12, 2014. As of June 1, 2015:
http://news.xinhuanet.com/english/china/2014-11/12/c_133785465.htm

Zetter, Kim, "The NSA Acknowledges What We All Feared: Iran Learns From U.S. Cyberattacks," *Wired*, February 10, 2015. As of February 20, 2015:
http://www.wired.com/2015/02/
nsa-acknowledges-feared-iran-learns-us-cyberattacks/